EARTHWORMS - THE WASTE MANAGERS: THEIR ROLE IN SUSTAINABLE WASTE MANAGEMENT CONVERTING WASTE INTO RESOURCE WHILE REDUCING GREENHOUSE GASES

WASTE AND WASTE MANAGEMENT

Additional books in this series can be found on Nova's website
under the Series tab.

Additional E-books in this series can be found on Nova's website
under the E-book tab.

WASTE AND WASTE MANAGEMENT

EARTHWORMS - THE WASTE MANAGERS: THEIR ROLE IN SUSTAINABLE WASTE MANAGEMENT CONVERTING WASTE INTO RESOURCE WHILE REDUCING GREENHOUSE GASES

RAJIV K. SINHA
SUNIL HERAT
SUNITA AGARWAL
KRUNAL CHAUHAN
DALSUKH VALANI

Nova Science Publishers, Inc.
New York

NOTICE TO THE READER

The Publisher has taken reasonable care in the preparation of this book, but makes no expressed or implied warranty of any kind and assumes no responsibility for any errors or omissions. No liability is assumed for incidental or consequential damages in connection with or arising out of information contained in this book. The Publisher shall not be liable for any special, consequential, or exemplary damages resulting, in whole or in part, from the readers' use of, or reliance upon, this material. Any parts of this book based on government reports are so indicated and copyright is claimed for those parts to the extent applicable to compilations of such works.

Independent verification should be sought for any data, advice or recommendations contained in this book. In addition, no responsibility is assumed by the publisher for any injury and/or damage to persons or property arising from any methods, products, instructions, ideas or otherwise contained in this publication.

This publication is designed to provide accurate and authoritative information with regard to the subject matter covered herein. It is sold with the clear understanding that the Publisher is not engaged in rendering legal or any other professional services. If legal or any other expert assistance is required, the services of a competent person should be sought. FROM A DECLARATION OF PARTICIPANTS JOINTLY ADOPTED BY A COMMITTEE OF THE AMERICAN BAR ASSOCIATION AND A COMMITTEE OF PUBLISHERS.

Additional color graphics may be available in the e-book version of this book.

LIBRARY OF CONGRESS CATALOGING-IN-PUBLICATION DATA

Earthworms - the waste managers : their role in sustainable waste management converting waste into resource while reducing greenhouse gases / authors: Rajiv K. Sinha ... [et al.].
   p. cm.
 Includes index.
  ISBN 978-1-61122-136-7 (softcover)
  1. Vermicomposting. 2. Earthworm culture. 3. Earthworms. 4. Organic wastes--Recycling. I. Sinha, Rajiv K. (Rajiv Kumar)
  TD796.5.E175 2011
  628.4'458--dc22
                          2010037363

Published by Nova Science Publishers, Inc. † New York

CONTENTS

Index

PREFACE

Millions of tons of municipal solid wastes (MSW) comprising a great proportion of 'food & green garden wastes' generated from homes & institutions are ending up in the landfills everyday, creating extraordinary economic problems for the local government & environmental problems for the society due to increasingly high cost of landfill construction, waste disposal & monitoring for emission of powerful 'greenhouse gases', 'toxic gases' and discharge of 'leachate' with serious risk of polluting groundwater.

Studies indicate that vermi-composting of food & garden wastes by waste-eater earthworms is more efficient over the conventional aerobic systems and the Tiger Worms (*Eisinea fetida*) are most voracious waste eaters and biodegraders. Vermicomposting is faster by 60-80%, reduces emission of GHG and the end-product is more nutritive, disinfected & detoxified. In the vermicomposting system 'with worms' degradation of mixed food wastes started within hours (5% after 24 hours) and after 15 days there was 100% degradation. But there was no degradation in the conventional aerobic system 'without worms' until after 30 days. This was achieved with about 1000 starter worms which multiplied to about 2500 worms during the 3 months of vermicomposting study.

Among the individual food components, bread, boiled rice and noodles, boiled potato, baked beans, green beans, pumpkin, tomato and lettuce leaves are degraded 50 - 100% in 'hours' (48 – 96 hours) by earthworms. They are among the 'preferred foods'. Pulses, pasta, orange and banana peels, lemons, broccoli, okra beans, watermelon skin, corn cobs, tea bags and tissue papers are degraded 50 - 100% in 7 – 14 days. In the system without worms 'none' or 'very poor' (max. 20%) degradation of these food components occurs in the same period of time.

Chapter 1

INTRODUCTION

Waste is a problem of the modern civilized society. We are facing the escalating economic and environmental cost of dealing with current and future generation of mounting municipal solid wastes (MSW). MSW is a term used to represent all the garbage created by households, commercial sites (restaurants, grocery and other stores, offices and public places etc.) and institutions (educational establishments, museums etc.). This also includes wastes from small and medium sized cottage industries. A considerable portion of this MSW consist of 'Organic Wastes' primarily 'food and garden wastes' which are 'biodegradable' and can be composted.

Millions of tons of MSW generated from the modern society are ending up in the landfills everyday, creating extraordinary economic and environmental problems for the local government to manage and monitor them (may be up to 30 years) for environmental safety (emission of greenhouse & toxic gases and leachate discharge threatening ground water contamination).(Lisk, 1991). Construction of secured engineered landfills incurs 20-25 million U.S. dollars before the first load of waste is dumped. In 2002-03 Australians generated 32.3 million tonnes of MSW of which 17.4 mt i.e. about 54% ended up in landfills (ABS, 2004).

Another serious cause of concern today is the emission of powerful greenhouse gases (GHG) methane (CH_4) & nitrous oxides (N_2O) resulting from the disposal of MSW either in the landfills or from their management by composting (Lou & Nair, 2009). Studies indicate high emissions of methane (CH_4) and nitrous oxide (N_2O) in proportion to the amount of food waste used. (Wang et al., 1997; Yaowu et al., 2000). Molecule to molecule CH_4 is 21 times and N_2O is 310 times more powerful GHG than the CO_2. AGO (2007) reported that disposal of MSW (primarily in landfills) contributed 17 million tonnes CO_2-e of GHG in Australia in 2005, equivalent to the emissions from 4 millions cars or 2.6% of the national GHG emissions.

Today community wastes are no longer seen as 'discarded materials' to be thrown away but as 'potential resource' to be recycled back into the human ecosystem for societal use. The 3 R's philosophy (reduction, reuse and recycling) of waste management is being advocated world wide for sustainable management of wastes for 'conservation of resources' (when waste is reduced & reused at source by society) as well as 'recovery of resources' (when waste materials are recycled in industries or at home to get new materials). This also leads to significantly reduced emission of greenhouse gases (Sinha, 2000; Sinha et al., 2008 a; Sinha & Bharambe, 2008)

Composting (biological recycling) of all 'organic wastes' (which are biodegradable) is emerging as a new tool in all developed world to divert large portion of community wastes (45 – 50%) from landfills. It is like 'killing two birds in one shot' - recovering 'nutritive fertilizer' while 'reducing' the needs of costly 'landfills'. (Gaur & Singh, 1995; Sinha et al., 2009 (a & b). It is in fact revival of the traditional systems practiced by farmers and ancient people in a more scientific way. In Australia and all other developed nations residents are be educated and motivated to compost all their food and garden wastes at home and divert them from ending up in landfills.

Composting of organic wastes by identified waste-eater composting earthworms (vermi-composting) is emerging as more efficient method over the conventional composting systems with multiple benefits to both environment and society (Datar et al., 1997 & Sherman, 2000). Vermi-composting is self-promoted, self-regulated, self-improved & self-enhanced, low or no-energy requiring zero-waste technology, easy to construct, operate and maintain. It excels all other composting (bio-degradation & bio-conversion of organic wastes) technologies by the fact that it can utilize waste organics that otherwise cannot be utilized by others; achieve greater utilization (rather than the destruction) of materials that cannot be achieved by others;

and that it does all with 'enzymatic actions'. (Dominguez, 1997 & Hand, 1998). Vermicomposting involves about 100-1000 times higher 'value addition' than other biological composting technologies (Appelhof, 2003).

Chapter 2

VERMICOMPOSTING OF ORGANIC WASTES : AN EFFICIENT & HYGIENIC SYSTEM

A revolution is unfolding in vermiculture studies (rearing of useful waste eater earthworms species) for rapid and odourless management of total municipal & industrial organic wastes in a much economically viable, socially acceptable and environmentally sustainable way (Dominguez, 2004). Earthworms can physically handle a wide variety of organic wastes (both solid and liquid) from both municipal (domestic and commercial) and industrial (livestock, food processing and paper industries) streams. They are highly adaptable to different types of organic wastes (even of industrial origin), provided the physical structure, pH and the salt concentrations are not above the tolerance level. Another matter of considerable significance is that the earthworms also partially 'detoxify' (by bio-accumulating any heavy metals and toxic chemicals) and 'disinfect' (by devouring on pathogens and killing them by anti-bacterial coelomic fluid) the waste biomass while degrading them into vermi-compost which is nearly chemical & pathogen free and odorless (Pierre et al., 1982).

Most earthworms consume, at the best, half their body weight of organics in the waste in a day. *Eisenia fetida* is reported to consume organic matter at the rate equal to their body weight every day (Visvanathan et al., 2005). Earthworm participation enhances natural biodegradation and decomposition of organic waste from 60 to 80%. The quality of compost is significantly improved in terms of nutritional & storage value by worms. (Klein et al., 2005).

WASTE EATER EARTHWORMS SUITABLE FOR WASTE DEGRADATION & STABILIZATION

Among the diverse earthworm species in the ecosystem some voracious waste-eaters have been identified (both exotic and local) and are now being used worldwide for domestic as well as commercial vermicomposting (Kaviraj & Sharma, 2003 & Nair, et al., 2007). Long-term researches into vermiculture have indicated that the Tiger Worm (*Eisenia fetida*), Red Tiger Worm (*E. andrei*), the Indian Blue Worm (*Perionyx excavatus*), the African Night Crawler (*Eudrilus euginae*), and the Red Worm (*Lumbricus rubellus*) are best suited for vermi-composting of variety of organic wastes (Graff, 1981; Reinecke et al., 1992 & Beetz, 1999). *E. fetida* and *E. andrei* are closely related. Tripathi & Bhardwaj (2004) also found *Lampito mauritii* as a good worm for composting kitchen wastes. Seenappa & Kale (1993), Seenappa et al., (1995), Kale & Sunitha (1995) & Lakshmi & Vizayalakshmi (2000) found *E. euginae* to be very suitable for composting some industrial organic wastes as well. Sinha et al., (2002 & 2005) studied vermicomposting of some domestic wastes with the mixed species of *E. fetida, P. excavatus* & *E. euginae* and came with very encouraging results. These species of worms are common to India and Australia, both nations being biogeographically very close. Reinecke et al., (1992) found these three species to work well even under higher temperatures. Our recent study has indicated that the Tiger Worm (*E. fetida*) is most versatile waste eater and degrader (Valani, 2009 & Chauhan, 2009). Garg et al., (2006) This voracious waste eater species is now being cultured on commercial scale in Australia by The Worm Man (www.thewormman.com) and sold in market to facilitate residents for vermicomposting their domestic wastes.

Chapter 3

WASTES SUITABLE FOR VERMICOMPOSTING

Waste eater earthworms can physically handle a wide variety of organic wastes from both municipal (domestic and commercial) and industrial (livestock, food processing and paper industries) streams. Earthworms are highly adaptable to different types of organic wastes (even of industrial origin), provided, the physical structure, pH and the salt concentrations are not above the tolerance level. (Garg et al., 2006).

MUNICIPAL ORGANIC WASTES

1. The food waste from homes (Some raw, but all cooked kitchen wastes - fruits and vegetables, grains & beans, coffee grounds, used tea leaves & bags, crushed egg shells) and restaurants & fried food wastes from fast-food outlets (Patil, 2005; Kristiana et al., 2005 & Suther, 2009; Valani, 2009; Chauhan, 2009);
2. The garden (yard) wastes (leaves and grass clippings) from homes and parks constitute an excellent feed stock for vermi-composting. Grass clippings (high carbon waste) require proper blending with nitrogenous wastes (Valani, 2009 & Chauhan, 2009);
3. The 'sewage sludge' (biosolids) from the municipal wastewater also provide a good feedstock for the worms. The worms digest the sludge and convert a good part of it into vermi-compost (Sinha et al., 2009 b);

4. Paunch waste materials (gut contents of slaughtered ruminants) from abattoir
5. also make good feedstock for earthworms (Fraser-Quick, 2002).

AGRICULTURE AND ANIMAL HUSBANDRY WASTES

1. Farm wastes such as crop residues, dry leaves & grasses;
2. Livestock rearing waste such as cattle dung, pig and chicken excreta makes excellent feedstock for earthworms (Edwards et al., 1985; Bansal & Kapoor, 2000);

(N.B. Animal excreta containing excessive nitrogen component may require mixing of carbon rich bulking agents e.g. straw, saw dust, dried leaves and grasses, shredded paper waste etc. to maintain proper C/N ratio).

SOME INDUSTRIAL ORGANIC WASTES SUITABLE FOR VERMI-COMPOSTING

Solid waste including the 'wastewater sludge' from paper pulp and cardboard industry, food processing industries including brewery and distillery; vegetable oil factory, potato and corn chips manufacturing industry, sugarcane industry, aromatic oil extraction industry. Sericulture industry, logging and carpentry industry also offers excellent feed material for vermi-composting by earthworms (Kale, et al., 1993; Kale & Sunitha, 1995; Seenappa & Kale, 1993; Seenappa et al., 1995; Gunathilagraj & Ravignanam, 1996; Lakshmi & Vijayalakshmi, 2000; Visvanathan et al., 2005).

WORMS CAN EVEN VERMICOMPOST HUMAN EXCRETA (FECES)

Eastman, (1999); Eastman et al., (2001) & Bajsa et al. (2004) reported that earthworms can even feed upon human excreta and convert them into vermicompost may be in longer time.

WORMS CAN FEED UPON MEAT WASTE PRODUCTS IF DRIVEN TO STARVATION

Our studies (Patil, 2005) found that worms can even eat chicken flesh (leaving the bones) if other feed materials are not available and driven to starvation. They are last food preferences. But the system is invaded with 'maggots' and some 'foul odour' for few days until worms eliminate them too by their ant-pathogenic actions.

Chapter 4

IMPORTANT STUDIES ON VERMICOMPOSTING OF WASTE

Vermicomposting of various types of organic wastes (municipal & industrial) has been successfully studied by several authors e.g. Kale, et al., (1993); Kale & Sunitha, (1995); Seenappa et al.,(1995); Gunathilagraj & Ravignanam, (1996); Elvira et al. (1998); Saxena et al. . (1998); Lotzof, (2000); Lakshmi & Vijayalakshmi (2000); Ndegwa & Thompson 2001; Fraser-Quick, 2002; Kaushik & Garg, 2000; Visvanathan et al., (2005); Bajsa et al. (2004);Contreras-Ramos et al. (2005) & Pramanik et al., (2007). Our current studies done at Griffith University, Australia has also discovered some new exciting facts about 'feeding behaviour & food preferences' of earthworms. Our studies also highlighted the 'greater efficiency of vermicomposting and better quality of compost' resulting from it as compared to the conventional 'aerobic' and 'anaerobic' composting systems. (Valani, 2009 & Chauhan, 2009; Sinha et al., 2009 a).

1. Kale, et al., (1993), Kale & Sunitha, (1995), Seenappa et al.,(1995), Gunathilagraj & Ravignanam, (1996), and Lakshmi & Vijayalakshmi (2000) studied the degradation and composting of 'wastewater sludge' from paper pulp and cardboard industry, brewery and distillery, sericulture industry, vegetable oil factory, potato and corn chips manufacturing industry, sugarcane industry, aromatic oil extraction industry, logging and carpentry industry by earthworms. These organic wastes offer excellent feed materials for vermi-composting by earthworms. Kale and Sunitha (1995) also studied the vermicomposting of waste from the mining industry which contains sulfur residues and creates disposal problems. They

can also be fed to the worms mixed with organic matter. Optimum mixing ratio of the sulfur waste residues to the organic matter was 4%.

2. Saxena et al. (1998) studied the vermicomposting of 'fly-ash' from the coal power plants which is considered as a hazardous waste and poses serious disposal problem due to heavy metal contents. As it is also rich in nitrogen and microbial biomass it can be vermi-composted by earthworms. They found that 25% of fly-ash mixed with sisal green pulp, parthenium and green grass cuttings formed excellent feed for Elsenia fetida and the vermi-compost was higher in NKP contents than other commercial manures. The earthworms ingest the heavy metals from the fly-ash while converting them into vermi-compost.

3. Sinha et al. (2002) studied the degradation and composting abilities of three species of earthworms (*Eisinea fetida, Eudrilus euginae & Perionyx excavatus*) on 'cattle dung and kitchen wastes' and found that although the worm *E. fetida* was a better waste degrader when used alone, an army of three species *E. fetida, E. euginae & P. excavatus* together degraded the waste faster.

4. Visvanathan et al., (2005) studied vermicomposting in greater details and concluded that most earthworms consume, at the best, half their body weight of organics in the waste in a day. *Eisenia fetida* can consume organic matter at the rate equal to their body weight every day. Earthworm participation enhances natural biodegradation and decomposition of organic waste from 60 to 80% over the conventional aerobic & anaerobic composting. Given the optimum conditions of temperature (20-30°C) and moisture (60-70%), about 5 kg of worms (numbering approx.10,000) can vermiprocess 1 ton of waste into vermi-compost in just 30 days. Upon vermi-composting the volume of solid waste is significantly reduced from approximately 1 cum to 0.5 cum of vermi-compost indicating 50% conversion rate, the rest is converted into worm biomass.

5. Contreras-Ramos et al. (2005) studied the vermicomposting of biosolids (dried sewage sludge) from various industries but mainly from textile industries and some households (municipal) mixed with cow manure and oat straw. 1,800, 1,400 and 1000 gms of aerobically digested biosolids were mixed with 800, 500 and 200 gms of cow manure and 200, 100 & 0 (zero) gms of oat straw in triplicate set up. A control was also kept with only biosolids. Cow

manure was added to provide additional nutrients and the oat straw to provide bulk. 50 earthworms (weighing 40 gm live weight) were added in each sample and the species used was *Eisenia fetida*. They were vermicomposted at three different moisture contents – 60%, 70% and 80% for two months (60 days). The best results were obtained with 1,800 g biosolids mixed with 800 g of cow manure and no (0) straw at 70% moisture content. Volatile solids of the vermicompost decreased by 5 times, heavy metals concentrations and pathogens (with no coliforms) were below the limits set by USEPA (1995) for an excellently stabilized biosolid. Carbon content decreased significantly due to mineralization of organic matter, and the number of earthworms increased by 1.2 fold.

6. Bajsa et al. (2004) studied vermicomposting of 'human excreta' (feces). It was completed in six months, with good physical texture meeting ARMANZ (1995) requirements, odourless and safe pathogen quality. Sawdust appeared to be the best covering material that can be used in vermicomposting toilets to produce compost with a good earthy smell, a crumbly texture and dark brown colour. There was no re-growth of pathogens on storing the compost for longer period of time and the initial pathogen load did not interfere in the die off process as the composting process itself seemed to stabilize the pathogen level in the system. They also studied the pathogen die-off in vermicomposting of sewage sludge spiked with *E.coli, S.typhimurium* and *E.faecalis* at the 1.6-5.4 x 10^6 CFU/g, 7.25 x 10^5 CFU/g and 3-4 x 10^4 CFU/g respectively. The composting was done with different bulking materials such as lawn clippings, sawdust, sand and sludge alone for a total period of 9 months to test the pathogen safety of the product for handling. It was observed that a safe product was achieved in 4-5 months of vermicomposting and the product remained the same quality without much reappearance of pathogens after in the remaining months of the test.

7. Pramanik et al., (2007) studied the vermicomposting of four (4) substrates viz. cow dung, grass, aquatic weeds and municipal solid wastes (MSW) to know the 'nutritional status & enzymatic activities' of the resulting vermicomposts in terms of increase in total nitrogen (N), total phosphorus (P) & potassium (K), humic acid contents and phosphatase activity. They found that cow dung recorded maximum increase in nitrogen (N) content (275%) followed by MSW (178%), grass (153%), and aquatic weed (146%)

in their resulting vermicomposts over the initial values in their raw materials. Application of lime however, increased N content in the vermicompost from 3% to 12% over non-limed treatment, irrespective of substrates used. Similarly, the vermicompost prepared from cow dung had the highest total phosphorus (12.70 mg/g) and total potassium (11.44 mg/g) over their initial substrate followed by those obtained from aquatic weeds, grasses and MSW. Humic acid was highest in vermicompost prepared from cow dung (0.7963 mg/g), followed by those from grasses (0.6147 mg/g), aquatic weeds (0.4724 mg/g) and MSW (0.3917 mg/g). Vermicompost obtained from cow dung showed the highest 'acid phosphatase' (200.45 µg p-nitrophenol /g/h) activities followed by vermicompost from grasses (179.24 µg p-nitrophenol /g/h), aquatic weeds (174.27 µg p-nitrophenol /g/h) and MSW (64.38 µg p-nitrophenol/g/h). The 'alkaline phosphatase' activity was highest in vermicompost obtained from aquatic weeds (679.88 µg p-nitrophenol/g/h) followed by cow dung (658.03 µg p-nitrophenol/g/h), grasses (583.28 µg p-nitrophenol/g/h) and MSW (267.54 µg p-nitrophenol/g/h).

8. Sinha et al., (2009 b) studied the biodegradation and vermicomposting of sewage sludge (biosolids) by earthworms. Earthworms feed readily upon the sludge components, rapidly convert them into vermicompost, reduce the pathogens to safe levels and ingest the heavy metals. Volume is significantly reduced from 1 cum of wet sludge (80% moisture) to 0.5 cum of vermicompost (30% moisture). Earthworms reduce the heavy metals, cadmium (Cd) and lead (Pb) from the digested sludge. There was no change in the values of heavy metals between the untreated sludge and those treated by adding cow dung and by organic soil enhance microbial composting. Although biosolids can be slowly stabilized by microbial degradation over a period of time, the heavy metals will remain in the system for quite sometimes after which it may leach into soil or get bound with soil organics by chemical reactions occurring in the soil. Providing additional feed materials to worms enhanced worm activity and also their number and led to greater removal of heavy metals. They inferred that over a period of time and with enhanced worm activity the heavy metals can be completely removed from the biosolids. They also found that the earthworms significantly reduce or almost eliminate the pathogens from the digested (composted) sludge. Sludge treated

with earthworms (with or without feed materials) only showed 'negative results' by the Colilert test under the UV lamp. And this was achieved in just 12 weeks. Study also inferred that under the conventional composting systems by enhanced microbial degradation the pathogens will remain in the biosolids for longer period of time until it is completely dry with all food and moisture exhausted making them difficult to survive.

Chapter 5

MECHANISM OF WORM ACTION IN THE COMPOSTING PROCESS

Worms degrade waste by 'multiple action'. First, grind the waste by 'muscular action' followed by breaking the waste organics by 'enzymatic action' and then by the action of 'decomposer microbes' which is proliferated by the earthworms in the composting system in billions & trillions.

A) WORMS DEGRADE WASTE BY GRINDING ACTION

Earthworms act as an aerator, grinder, crusher, chemical degrader and a biological stimulator in any environment they live and operate. (Dash, 1978; Binet et al., 1998 & Sinha et al., 2002). The waste feed materials ingested is finely ground (with the aid of stones in their muscular gizzard) into small particles to a size of 2-4 microns and passed on to the intestine for enzymatic actions. The gizzard and the intestine work as a 'bioreactor';

B) EARTHWORMS REINFORCE MICROBIAL POPULATION AND ACT SYNERGISTICALLY WITH MICROBES

Earthworms promotes the growth of 'beneficial decomposer microbes' (bacteria, actinomycetes & fungi) in waste biomass and this they do by improving aeration by burrowing actions. Earthworms hosts millions of decomposer (biodegrader) microbes in their gut which is described as 'little bacterial factory'. They devour on microbes and excrete them out (many times more in number than they ingest) in soil along with nutrients nitrogen (N) and phosphorus (P) in their excreta (Singleton et al., 2003). The nutrients N & P are further used by the microbes for multiplication and vigorous action. Edward and Fletcher (1988) showed that the number of bacteria and 'actinomycetes' contained in the ingested material increased up to 1000 fold while passing through the gut. A population of worms numbering about 15,000 will in turn foster a microbial population of billions of millions. (Morgan & Burrows, 1982). Singleton et al. (2003) studied the bacterial flora associated with the intestine and vermicasts of the earthworms and found species like *Pseudomonas, Mucor, Paenibacillus, Azoarcus, Burkholderia, Spiroplasm, Acaligenes,* and *Acidobacterium* which has potential to degrade several categories of organics. *Acaligenes* can even degrade PCBs and Mucor can degrade dieldrin.

Under favorable conditions, earthworms and microorganisms act 'symbiotically & synergistically' to accelerate and enhance the decomposition of the organic matter in the waste. It is the microorganisms which breaks down the cellulose in the food waste, grass clippings and the leaves from garden wastes (Morgan & Burrows, 1982; Xing et al. 2005).

C) WORMS DEGRADE WASTE BY ENZYMATIC ACTION

1. The worms secrete enzymes proteases, lipases, amylases, cellulases and chitinases in their gizzard and intestine which bring about rapid biochemical conversion of the cellulosic and the proteinaceous materials in the waste organics. Earthworms convert cellulose into its food value faster than proteins and other carbohydrates. They ingest the cellulose, pass it through its intestine, adjust the pH of the digested (degraded) materials, cull the unwanted microorganisms, and then deposit the processed cellulosic materials mixed with minerals and microbes as aggregates called 'vermicasts' in the soil. (Dash, 1978).

2. Only 5-10 percent of the chemically digested and ingested material is absorbed into the body and the rest is excreted out in the form of fine mucus coated granular aggregates - the 'vermicasts' rich in nitrates, phosphates and potash.

D) HUMIFICATION OF DEGRADED WASTE

The final process in vermi-processing and degradation of organic waste is the 'humification' in which the large organic particles are converted into a complex amorphous colloid containing 'phenolic' materials. Only about one-fourth of the organic matter is converted into humus. The colloidal humus acts as 'slow release fertilizer' (Visvanathan et al. 2005).

Chapter 6

CRITICAL FACTORS AFFECTING VERMICOMPOSTING BY EARTHWORMS

1. MODERATE TEMPERATURE

Vermicomposting is a 'mesophilic composting' system where temperature does not increase beyond 30°C. In general earthworms prefers and tolerates cold and moist conditions far better than the hot and dry ones. Most worms involved in vermi-composting require moderate temperature between 20 – 30°C. They are at the highest levels of both waste degradation and reproduction activity in the cool and warms weathers.

2. ADEQUATE MOISTURE CONTENT

Moisture is also a critical factor in vermicomposting process. It helps in the biochemical reaction and also retains heat. Moisture content of 60 - 70% of total weight of waste is considered to be ideal for vermicomposting (Edward & Lofty, 1972; Dynes, 2003). Our study (Valani, 2009; Chauhan, 2009) however, found that *Eisinea fetida* performed well around 50% moisture content. Worm body contain plenty of water which it uses during the period of water scarcity in the waste biomass.

3. ADEQUATE AERATION

Vermi-composting is an aerobic process and adequate flow of air in the waste biomass is essential for worm function. Worms breathe through their skin. However, in our recent study (Chauhan, 2009) we also discovered worms in 'anaerobic bins' devoid of aeration systems. No worms were originally added in the waste and they originated from the garden soil underneath. This may be due to the fact that worms constantly aerate their habitat by 'burrowing actions' and the amount of air which enters the bin everyday during the opening of the cover to dispose wastes is enough for the worms to survive. Again, the species was identified as *E. fetida* (Chauhan, 2009).

4. PH OF THE WASTE PILE

Earthworms are sensitive to pH change. Although they can survive in a pH range of 4.5 to 9 but functions best at neutral pH of 7.0 (Edwards, 1998). Also, to minimize the loss of nitrogen in the form of ammonia, pH should not go above 8.5. Organic materials are naturally well buffered relative to pH changes. But as composting progress, a slightly lower pH can be expected in the final products due to the formation of carbonic acid from the carbon dioxide produced by the breakdown of organics. Although worms can raise pH of its medium by secreting calcium (Ca), it is suggestive to add lime (calcium carbonate) powder to the waste biomass periodically. This serves two purposes - maintain neutral pH and also supply the much needed calcium (Ca) to the worms for its metabolism.

5. ADEQUATE AVAILABILITY OF CALCIUM (CA)

Calcium appears to be important mineral in worm biology (as calcarious tissues) and biodegradation activity. Although most organic waste contains calcium, it is important to add some additional sources of calcium e.g. lime powder for good vermi-composting. Pramanik et al., (2007) found that application of lime @ 5 gm/kg of substrate not only enhances the rate of vermicomposting but also results into nutritionally better vermicompost with greater enzymatic (phosphatase & urease) activities. It increased nitrogen

(N) content in the vermicompost from 3% to 12% over non-limed treatment. There was also increase in total phosphorus (11-22%) in the vermicompost after liming. It also increased humic acid contents and the population of total bacterial count, the cellulolytic fungi and the nitrogen-fixing bacteria. Our study (Chauhan, 2009 & Valani 2009) also indicate that addition of lime powder increases the efficiency of vermicomposting.

6. CARBON/NITROGEN (C/N) RATIO OF THE WASTE MATERIAL

Nitrogen is a 'critical factor' in any aerobic composting system. Generally 25 parts carbon to 1 part nitrogen by weight (C/N=25:1) is considered ideal for rapid composting. High C/N ratio above 30:1 in waste biomass has been found to impair worm activity and vermi-composting (Edwards & Bohlen, 1996). Although earthworms help to lower the C/N ratio of fresh organic waste, it is advisable to add nitrogen supplements such as cattle dung or pig and goat manure or even kitchen waste when waste materials of higher C/N ratio exceeding 40:1 such as the cellulosic green wastes are used for vermi-composting.

7. ADEQUATE NUMBER AND BIOMASS OF COMPOSTING WORMS

The number and quantity (biomass) of earthworms is also a 'critical factor' for vermi-composting. More the number of worms, rapid is the decomposition and also 'odor-free'. A minimum of about 100-150 adult worms per kg of waste in the initial stage is considered ideal. However, as the worms multiply at a rapid rate the number of worms quickly increases under ideal conditions of food, moisture and temperature and the degradation process becomes faster.

Chapter 7

EXPERIMENTAL STUDY OF COMPOSTING OF MIXED FOOD & GARDEN WASTES BY AEROBIC, ANAEROBIC AND VERMICOMPOSTING SYSTEMS IN METHODICAL & CASUAL WAYS

This study was designed to compare degradation rate of mixed food & garden wastes (mainly grass clipping) by aerobic, anaerobic & vermicomposting systems when done in 'methodical' way i.e. caring for adequate moisture of the waste biomass, aeration of compost pile, & addition of mulch (to create air pockets in aerobic bins), garden soil (to add decomposer microbes) & lime (to maintain neutral pH in the decomposing waste biomass for efficient microbial action) etc. time to time, and when done just in 'casual' way i.e. not caring much for adequate moisture, aeration of the aerobic bins, no further addition of lime and soil after first application. There are both types of people in our literate society, some take care about proper disposal of wastes, other take it casually.

MATERIALS AND METHODS

Composting bins of recycled plastics (220 L - 240 L capacity) with cover (lid) on top are available in hardware stores all over Australia. Aerobic & vermicomposting bins (worm farm) are ventilated from four sides for free flow of air, while the anaerobic bin is completely sealed from sides and bottom with no provision for ventilation from anywhere.

The bins were kept in glasshouse under shade to avoid direct sunlight. On the bottom of the bins about 6 cm thick layer of soil was spread to form the 'composting base'. On the layer of soil, about 3 kg of mixed food wastes were added. About 400 grams of grass clippings was spread on the food waste. In the Vermicomposting bin about 1000 composting earthworms (*Eisinea fetida*) were added as 'starter dose' and there was no further addition of worms. Red worms (*E. fetida*) is cultured & sold by 'The Worm Man' (www.thewormman.com) in Australia. Finally, a thin layer of garden soil was spread on the top of the waste biomass. This helps initiate decomposition of waste by decomposer microbes in soil.

Measured amount of mixed food & garden wastes were added every week and observations were made every fortnight. In all the methodical composting bins about 25 gm of 'lime' and about 50 gm of 'pine bark' was also added time to time. Water was sprinkled in these bins periodically to maintain adequate moisture level of around 60 – 70%. The aerobic and the vermicomposting bins were regularly aerated by manually turning the waste biomass.

Photo of Aerobic, Anaerobic and Vermicomposting bins.

(A1). Anaerobic Bin; (A2) Aerobic Bin; (A3) Vermicomposting Bin (Methodical Composting).

(B1). Anaerobic Bin; (B2) Aerobic Bin; (B3) Vermicomposting Bin (Casual Composting).

Aerobic Composting (Day 1)

Aerobic Composting (Day 90)

Anaerobic Composting (Day 1)

Anaerobic Composting (Day 90)

Vermicomposting (Day 1)

Vermicomposting (Day 90)

Photo showing degradation of mixed food and garden wastes in aerobic, anaerobic & vermicomposting bins in 90 days period when done 'methodically'.

Table 1. Percentage degradation of mixed food wastes & garden wastes (grass clippings) by vermicomposting vis-à-vis aerobic & anaerobic composting systems in methodical and casual ways

Waste Materials	Vermicomposting By Earthworms (%)		Aerobic Composting (%)		Anaerobic Composting (%)	
	M	C	M	C	M	C
1. Mixed Food Waste						
After 24 hours	5%	0	0	0	0	0
After 15 Days	100%	60	0	0	0	0
After 30 Days	100	60	0	0	0	0
After 45 Days	100	70	5	0	10	0
After 60 Days	100	70	25	10	40	40
After 75 Days	100	80	30	15	5	70
After 90 Days	100	90	35	20	85	75
2. Garden Waste						
After 15 Days	15%	15	0	0	0	0
After 30 Days	35	15	0	0	0	0
After 45 Days	40	18	0	0	5	0
After 60 Days	100%	70	15	10	50	50
After 75 Days	100	70	20	10	70	40
After 90 Days	100	80	35	12	80	60

Keys: M = Methodical; C = Casual.

OBSERVATIONS, RESULTS & DISCUSSION

VERMICOMPOSTING systems both 'methodical' as well as 'casual' excelled over the conventional aerobic and anaerobic composting systems in degrading food & garden wastes. In the methodical vermicomposting system degradation of mixed food wastes had started within hours (5% after 24 hours) and after 15 days there was 100% degradation. Even in the casual vermicomposting there was 60% degradation but none in the conventional aerobic and anaerobic systems until after 30 days. At the end of the study period i.e. after 90 days, even the casual vermicomposting systems proved more efficient over the methodical aerobic & anaerobic systems. And these results were obtained with about 1000 to 1500 worms in the vermicomposting bin.

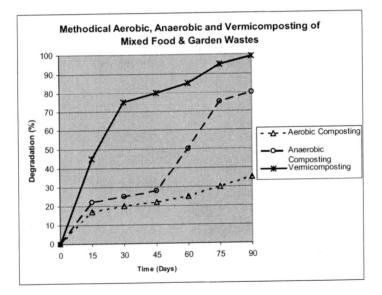

Figure 1. Comparison of the three composting systems of mixed food & garden wastes when done methodically.

However, among the conventional composting systems, the anaerobic system performed more efficiently over the aerobic systems disproving the general belief that aerobic composting is faster than anaerobic. It degraded over 60 – 80% of the wastes in the casual & methodical methods respectively, against maximum 35% by the aerobic system.

In all systems, FOOD WASTES were 'faster' to degrade than the GARDEN WASTES. Whereas, food wastes were degraded 100% in just 15 days, it took over 60 days for the garden wastes to degrade by 100%. There was some odour problem in the anaerobic system at the time of addition of waste, but other two systems were odour free.

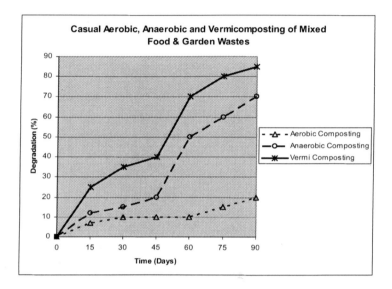

Figure 2. Comparison of the three composting systems of mixed food & garden wastes when done casually.

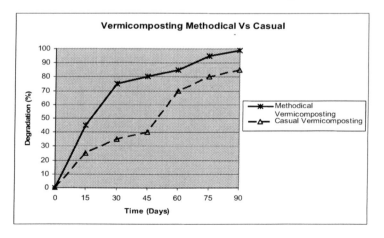

Figure 3. Comparison of Degradation of Mixed Food & Garden Wastes by Methodical Vs Casual Vermicomposting Systems.

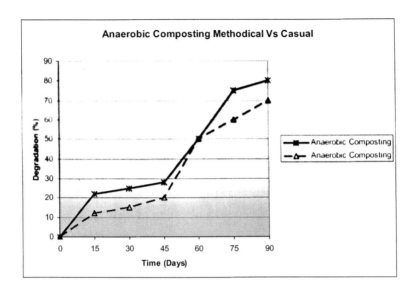

Figure 4. Comparison of Degradation of Mixed Food & Garden Wastes by
Methodical Vs Casual Anaerobic Composting System.

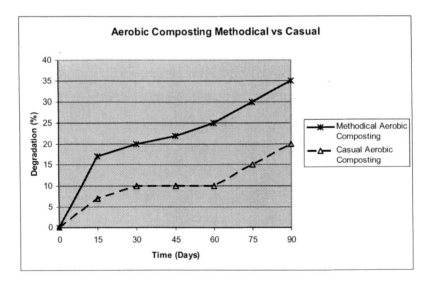

Figure 5. Comparison of Degradation of Mixed Food & Garden Wastes by
Methodical Vs Casual Aerobic Composting System.

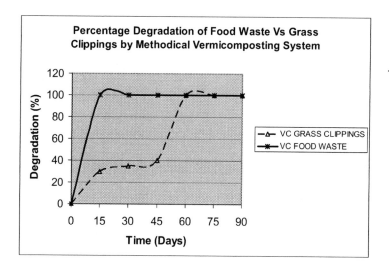

Figure 6. Comparison of Degradation of Food Waste Vs Grass Clippings by Methodical Vermicomposting System.

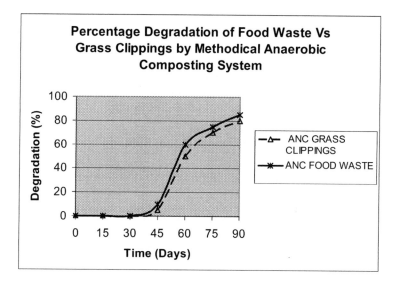

Figure 7. Comparison of Degradation of Food Waste Vs Grass Clippings by Methodical Anaerobic Composting System.

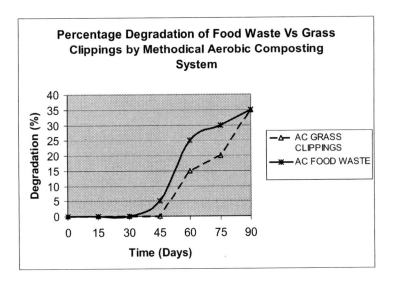

Figure 8. Comparison of Degradation of Food Waste Vs Grass Clippings by
Methodical Aerobic Composting System.

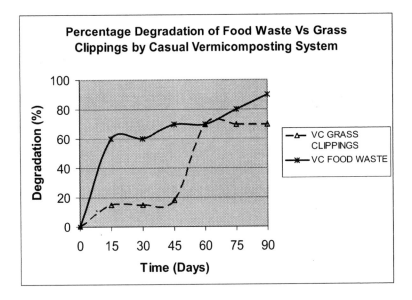

Figure 9. Comparison of Degradation of Food Waste Vs Grass Clippings by Casual
Vermicomposting System.

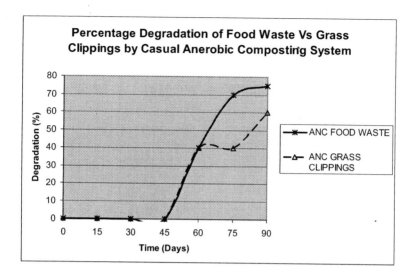

Figure 10. Comparison of Degradation of Food Waste Vs Grass Clippings by Casual Anaerobic Composting System.

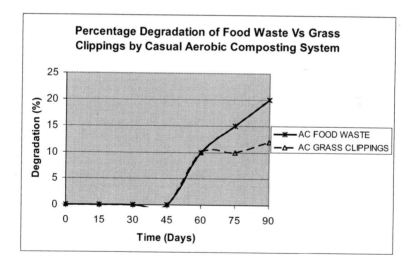

Figure 11. Comparison of Degradation of Food Waste Vs Grass Clippings by Casual Aerobic Composting System.

Chapter 8

EXPERIMENTAL STUDY OF COMPOSTING OF INDIVIDUAL FOOD & GARDEN WASTE COMPONENTS BY AEROBIC, ANAEROBIC AND VERMICOMPOSTING SYSTEMS IN METHODICAL & CASUAL WAYS

This study was designed to compare degradation rate of individual food & garden wastes components listed in table-2 by aerobic, anaerobic & vermicomposting systems when done in both 'methodical' and 'casual' ways.

Table 2. Percentage degradation of individual food waste components by vermicomposting vis-à-vis aerobic & anaerobic composting systems in methodical and casual ways

Food & Garden Waste Component (In hours & days)		Vermicomposting (%)		Aerobic Composting (%)		Anaerobic Composting (%)	
		M	C	M	C	M	C
1	Bread Slice (48 h)	100%	0	0	0	0	0
2	Raw Tomato (72 h)	70	65	20	20	10	5
3	Pumpkin Cut (72 h)	50	30	0	0	0	0
4	Lettuce Leaf (72 h)	50	10	0	0	20	0
5	Boiled Potatoes (72 h)	50	10	0	0	0	0
6	Indian Baked Bread (4 d)	100%	40	5	5	5	5

Table 2. (Continued).

Food & Garden Waste Component (In hours & days)		Vermicomposting (%)		Aerobic Composting (%)		Anaerobic Composting (%)	
		M	C	M	C	M	C
7	Boiled Noodles (4 d)	100%	80	20	5	30	10
8	Baked Beans (4 d)	90	5	0	0	0	0
9	Cooked Rice (4 d)	70	0	0	0	0	0
10	Green Beans (4 d)	60	0	5	20	0	5
11	Lemon Squeezed (4 d)	50	30	0	5	0	0
12	Boiled Pasta (7 d)	100%	60	10	10	10	10
13	Tissue Paper (7 d)	99%	20	0	0	10	70
14	Okra Beans (7 d)	90	60	20	40	5	5
15	Boiled Pulse (7 d)	60	35	20	10	60	40
16	Banana Peels (7 d)	50	30	40	30	15	10
17	Watermelon Skin (7 d)	100%	99	75	90	95	90
18	Indian Fried Bread (11 d)	100%	10	10	10	5	5
19	Broccoli Cuts (11 d)	75	30	10	10	60	50
20	Corn Cob (11 d)	50	10	10	10	10	10
21	Orange Peels (14 d)	100%	60	20	30	20	30
22	Tea Bags (14 d)	90	50	10	0	40	30
23	Garden Leaves (11 d)	70	30	20	0	70	40
24	Grass Clippings (14 d)	40	20	10	5	30	30
25	Paper Towel (14 d)	20	0	0	0	30	10
26	Raw Potato (14 d)	10	5	5	5	10	15
27	Raw Onion (14 d)	10	10	15	5	15	10
28	Crushed Egg Shell (14 d)	5%	0	0	0	0	0
29	Corn Peel (14 d)	0	0	0	0	0	0

Keys:
M = Methodical;
C = Casual;
h = hours;
d = days.

(A). Photo showing different degree of degradation of some individual food components by vermicomposting, anaerobic and aerobic composting systems.

Photo. Individual Food Components (Day 1) Vermicomposted by Earthworms (Day 15).

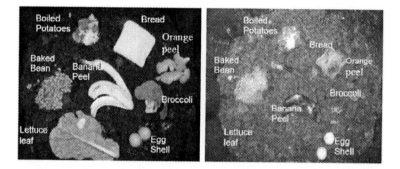

Photo. Individual Food Components (Day 1) Composted by Anaerobic System (Day 15).

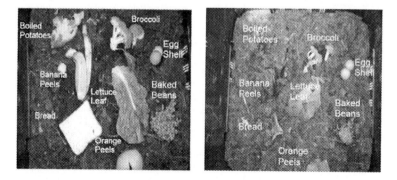

Photo. Individual Food Components (Day 1) Composted by Aerobic System (Day 15).

(B). Photo showing different degree of degradation of some individual food components by vermicomposting, anaerobic and aerobic composting systems.

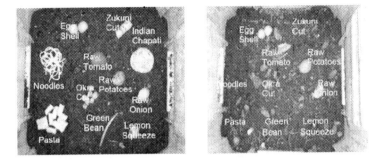

Photo. Individual Food Components (Day 1) Vermicomposted by Earthworms (Day 15).

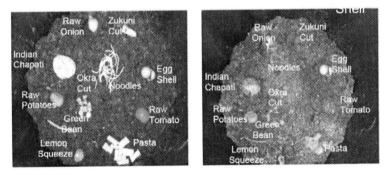

Photo. Individual Food Components (Day 1) Composted by Anaerobic System (Day 15).

Photo. Individual Food Components (Day 1) Composted by Aerobic System (Day 15).

(C). Photo showing different degree of degradation of some individual food components, tissue papers and paper towels by vermicomposting, anaerobic and aerobic composting systems.

Photo. Individual Food Components (Day 1) Vermicomposted by Earthworms (Day 15).

Photo. Individual Food Components (Day 1) Composted by Anaerobic System (Day 15).

Photo. Individual Food Components (Day 1) Composted by Aerobic System (Day 15).

(D). Photo showing different degree of degradation of some individual food components, garden waste (leaves and grass clippings) by vermicomposting, anaerobic and aerobic composting systems.

Photo. Individual Food Components (Day 1) Vermicomposted by Earthworms (Day 15).

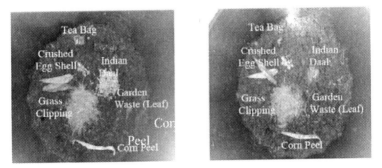

Photo. Individual Food Components (Day 1) Composted by Anaerobic System (Day 15).

Photo. Individual Food Components (Day 1) Composted by Aerobic System (Day 15).

RESULTS AND DISCUSSION ON INDIVIDUAL FOOD & GARDEN WASTE COMPOSTING

It can be seen that bread, boiled potato, baked beans, pumpkin, tomato & lettuce leaves which constitute an important component & large volume in all food wastes in the multicultural Australian society is degraded & composted between 50 to 100% in HOURS by earthworms. Baked food products such as bread & buns specially constitute large volumes in food wastes. Boiled rice & pulses, pasta & noodles, orange & banana peels, lemons, green beans, broccoli, okra beans, watermelon skin, corn cobs & tea bags also constitute large volumes in food wastes as they are universally consumed. Tissue papers are used and thrown in large amounts in every home every day. They are all degraded between 50 to 100% by earthworms in few DAYS to few WEEKS. Similar should be the fates of other food waste components e.g. carrots, pumpkin, capsicum, mushroom, celery, apple, pear, grapes and kiwi fruits consumed in large quantities in modern societies. And because these results were obtained with about 1000 to 1500 worms in the vermicomposting bin and as the number of worms multiply rapidly and vermi-degradation process becomes faster with time, they can be presumed to be 100% composted within 3 to 4 weeks.

But concern remains about the degradation of 'raw potatoes & onions', 'corn peels' and the 'egg shells' which are also rejected and thrown as waste in large amounts in all societies. They will take longer time in degradation and composting.

After the vermicomposting system, it is again the 'anaerobic system' which appears to handle several of the food & garden wastes components (grass clippings & leaves of flowering plants) more efficiently.

Chapter 9

ADVANTAGES OF VERMICOMPOSTING OVER CONVENTIONAL COMPOSTING

Earthworms have real potential to both increase the rate of aerobic decomposition and composting of organic matter and also to stabilize the organic residues in the waste removing the harmful pathogens and heavy metals (if any) from the compost. The quality of compost is significantly better than conventional composts, more homogenous, rich in key minerals & beneficial soil microbes. In fact in the conventional composting systems which is thermophilic (temperature rising up to 55°C) many beneficial microbes are killed and some nutrients especially nitrogen (N) is lost (due to gassing off of nitrogen).

1. NEARLY ODOR-FREE PROCESS

Earthworms inhibit the action of anaerobic micro-organisms in waste (due to aeration by burrowing actions) which release foul-smelling hydrogen sulfide and mercaptans.

2. EARTHWORMS MINERALIZE THE NUTRIENTS FROM WASTE ORGANICS TO INCREASES ITS VALUE IN THE END PRODUCT & MAKE THEM BIO-AVAILABLE

Earthworms induces several beneficial changes in the biochemical properties of the composting wastes (Atiyeh et al., 2000). Most important is that they mineralize the nitrogen (N), phosphorus (P), potassium (K) and all other nutrients in the waste organics to 'increase' their value in the end products (vermicompost) and also make them 'bio-available' to plants. (Buchanan et al., 1988). They ingest nitrogen from the waste and excrete it in the mineral form as nitrates, ammonium and muco-proteins. The nitrogenous waste excreted by the nephridia of the worms is plant-available as it is mostly urea and ammonia. The ammonium in the soil is bio-transformed into nitrates. Phosphorus (P) contents increased in the vermicomposted waste treated with earthworms but decreased in the samples without worms (Parvaresh, et al., 2004).

Pramanik et al., (2007) studied the vermicomposting of cow dung and municipal solid wastes and found that nitrogen (N) content in the composted cow dung increased by 275% followed by MSW (178%). Similarly, the vermicompost prepared from cow dung increased phosphorus (12.70 mg / g) and potassium (11.44 mg / g) over their initial raw materials.

Our study (Agarwal, 1999) found that the NPK contents of the vermicomposted cattle dung increased significantly as compared to the conventionally composted cattle dung. The NPK in conventional cattle dung compost ranged between 0.4-1.0%, 0.4-0.8% & 0.8 -1.2% respectively, while in the vermicompost the NPK ranged between 2.5-3.0%, 1.8-2.9% & 1.4-2.0% respectively. Our other study (Singh, 2009) also found that nitrogen (N) content in the vermicompost was 9.5 mg/L while it was only 6 mg/L and 5.7 mg/L respectively in the conventional aerobic and anaerobic composts made from the same feedstock – food and garden wastes. Similar was the situation with other nutrients like phosphorus (P), potassium (K), magnesium (Mg), manganese (Mn) and calcium (Ca). They were significantly higher in vermicompost.

3. EARTHWORMS PROLIFERATE BENEFICIAL MICROBES IN THE END PRODUCT

Studies indicate that composts made by worm action are rich in 'microbial populations & diversity', particularly 'fungi', 'bacteria' and 'actinomycetes'. Worms have been found to particularly proliferate *Actinomycetes, Azotobacter, Rhizobium, Nitrobacter* & phosphate solubilizing bacteria significantly in their end products. (Suhane, 2007; Pramanik, 2007).

Our study (Singh, 2009) also indicated higher values of '*Azotobacter*' (the nitrogen fixing bacteria) and the '*Actinomycetes*' (the bacteria that increases biological resistance in pants against pests and diseases) in vermicompost as compared to the conventional aerobic and anaerobic composts.

4. EARTHWORMS DESTROY HARMFUL MICROBES IN THE END PRODUCT MAKING THEM PATHOGEN FREE & HYGIENIC

The earthworms release coelomic fluids that have anti-bacterial properties and destroy all pathogens in the waste biomass. (Pierre et al., 1982). They also devour the harmful protozoa, bacteria and fungus as food. They seems to realize instinctively that anaerobic bacteria and fungi are undesirable and so feed upon them preferentially, thus arresting their proliferation. More recently, Dr. Elaine Ingham has found in her research that worms living in pathogen-rich materials (e.g. sewage and sludge), when dissected, show no evidence of pathogens beyond 5 mm of their gut. This confirms that something inside the worms destroys the pathogens, and excreta (vermicast) becomes pathogen-free (Appelhof, 2003). Eastman et al., (2001) also reported significant pathogen reduction by vermicomposting.

In the intestine of earthworms some bacteria & fungus (*Pencillium* and *Aspergillus*) have also been found (Singelton et al., 2003). They produce 'antibiotics' and kills the pathogenic organisms in the sewage sludge making it virtually sterile. The removal of pathogens, faecal coliforms (*E. coli*), *Salmonella* spp., enteric viruses and helminth ova appear to be much more rapid when they are processed by *E. fetida*. Of all *E. coli* and *Salmonella* are greatly reduced (Bajsa et al., 2004).

Our study (Singh, 2009) also found that there were significantly much lesser numbers of coliform colonies in vermicompost as compared to the conventionally produced aerobic and anaerobic compost. This indicates that vermicompost is more safe and hygienic than the aerobic and anaerobic composts although prepared from same raw materials.

Studies by Standard Australia (2003) suggest thermophilic composting of pathogenic raw materials like 'sewage sludge' before vermicomposting for complete elimination of pathogens. Nair et al., (2006) also found that 9 days pre-composting of kitchen wastes (food scraps) followed by vermicomposting eliminates the pathogenic microbes including coliforms more effectively from the vermicompost.

5. EARTHWORMS REMOVES TOXIC CHEMICALS FROM COMPOSTED PRODUCTS

Several studies have found that earthworms effectively bio-accumulate or biodegrade several organic and inorganic chemicals including 'heavy metals', 'organochlorine pesticide' and 'polycyclic aromatic hydrocarbons' (PAHs) residues in the medium in which it feeds. (Ireland, 1983 & Sinha et al., 2008 a).

6. LOWER GREENHOUSE GAS EMISSIONS BY VERMI-COMPOSTING OF WASTE

Emission of powerful greenhouse gases methane (CH_4) and nitrous oxide (N_2O) in waste management programs of both municipal solid wastes (MSW) & sewage has become a major global issue today in the wake of increasing visible impacts of global warming. Biodegradation of organic waste has long been known to generate methane (CH_4). Studies have also indicated high emissions of nitrous oxide (N_2O) in proportion to the amount of food waste used, and methane (CH_4) is also emitted if the composting piles contain cattle manure. (Toms et al. 1995; Wu et al. 1995; Wang et al., 1997; & Yaowu et al., 2000).

· High volumes of carbon dioxide (CO_2) and methane (CH_4) is emitted from the conventional composting process especially in anaerobic conditions. Our study found that on average the anaerobic composting systems emitted the highest amount of CO_2 (2950 mg/m^2/hour) and CH_4

(9.54 mg/m^2/hour), while the aerobic systems (both with and without worms) emitted the least amount of CO_2 (880 mg/m^2/hour) and CH_4 (2.17 mg/m^2/hour). Vermicomposting systems had the 'lowest emission' of N_2O which is most powerful GHG (Sinha & Chan, 2009).

In theory, vermicomposting by worms should provide some potentially significant advantages over conventional composting with respect to emissions of methane (CH_4). Worms significantly increase the proportion of 'aerobic to anaerobic decomposition' in the compost pile by burrowing and aerating actions leaving very few anaerobic areas in the pile, and thus resulting in a significant decrease in methane (CH_4) and also volatile sulfur compounds which are readily emitted from the conventional (microbial) composting process (Mitchell et al., 1980). Analysis of vermicompost samples has shown generally higher levels of available nitrogen (N) as compared to the conventional compost samples made from similar feedstock. This implies that the vermicomposting process by worms is more efficient at retaining nitrogen (N) rather than releasing it as N_2O.

7. No or Low Energy Use in Vermi-composting Process

Conventional microbial composting requires energy for aeration (constant turning of waste biomass and even for mechanical airflow) and sometimes for mechanical crushing of waste to achieve uniform particle size. Vermi-composting do not involve such use of energy.

Table 3. GHG emission rates in aerobic, anaerobic and vermicomposting bins (min - max values in bracket)

Greenhouse Gases	Aerobic bins	Anaerobic bins	Vermicomposting bins
Number of samples	40	45	42
GHG Emissions (mg m^{-2} hr^{-1})			
CO_2	882 (23 – 5764)	2950 (91 – 10069)	1675 (146 – 5669)
CH_4	2.17 (0.00 – 38.05)	9.54 (0.00 – 52.90)	4.76 (0.00 – 40.89)
N_2O	1.48 (0.01 – 16.25)	1.59 (0.00 – 16.37)	1.17 (0.00 – 24.78)
Total Emissions (mg CO_2-e m^{-2} hr^{-1})			
Excluding CO_2	504 (4 – 5038)	694 (0.76 – 5073)	463 (4 – 8475)
Including CO_2	1386 (28 – 7554)	3644 (259 – 14351)	2138 (189 – 14144)

Source: Sinha and Chan (2009).

8. EARTHWORMS BIOMASS: A SOCIALLY AND ECONOMICALLY VALUABLE BY-PRODUCT OF VERMICOMPOSTING

Huge population of earthworms results from vermicomposting of wastes. They are good source of nutritive 'worm meal' rich in proteins (65%) with 70-80% high quality essential amino acids 'lysine' and 'methionine'. It is superior to even 'fish meal' and 'meat meal' and a wonderful feed materials for fish, cattle and poultry (Sabine, 1978).

In the last 10 years, a number of earthworm's 'clot-dissolving', 'lytic' and 'immune boosting' compounds have been isolated and tested clinically. Current researches made in Canada, China, Philippines, Japan and other countries on the identification, isolation and synthesis of some 'bioactive compounds' from earthworms (*L. rubellus* and *E. fetida*) with potential medicinal values have brought revolution in the vermiculture studies (Wang 2000). Oral administration of earthworms powder & enzymes were found to be effective in treating 'thrombotic diseases', 'arthritis', 'diabetes mellitus', 'pulmonary heart disease', 'lowering blood pressure', 'epilepsy', 'schizophrenia', 'mumps', 'exzema', 'chronic lumbago', 'anemia', 'vertigo' and 'digestive ulcer' (Mihara et al. 1990; Wengling and Sun 2000; Lopez and Alis 2005).

Some biological compounds from earthworms are also finding new industrial applications. Stearic acid found in earthworms is a long chain saturated fatty acid and are widely used as 'lubricant' and as an 'additive' in industrial preparations. It is used in the manufacture of metallic stearates, pharmaceuticals soaps, cosmetics and food packaging. It is also used as a 'softner', 'accelerator activator' and 'dispersing agents' in rubbers. Industrial applications of lauric acid and its derivatives are as 'alkyd resins', 'wetting agents', a 'rubber accelerator' and 'softner' and in the manufacture of 'detergents' and 'insecticides' (Kangmin 1998; Lopez & Alis, 2005).

Chapter 10

WASTE MANAGEMENT BY EARTHWORMS: A GLOBAL MOVEMENT

The movement was started in the middle of 20th century and the first serious experiments for management of municipal/industrial organic wastes were established in Holland in 1970, and subsequently in England, and Canada. Later vermiculture were followed in USA, Italy, Philippines, Thailand, China, Korea, Japan, Brazil, France, Australia and Israel (Edward, 1988). However, the farmers all over the world have been using worms for composting their farm waste and improving farm soil fertility since long time.

1) U.S.A.

U.S. has some largest vermicomposting companies and plants in world and states are encouraging people for 'backyard vermicomposting' to divert wastes from landfills (Bogdanov, 1996 & 2004). The American Earthworm Company started a 'vermi-composting farm' in 1978-79 with 500 t month of vermicompost production (Edwards, 2000). A farm in LA rears 1,000,000 worms to treat 7.5 tons of garbage each month. Nearly 300 large-scale vermiculturist formed an 'International Worms Growers Association' in 1997 and is having booming business. Vermicycle Organics produced 7.5 million pounds of vermicompost every year in high-tech greenhouses. Its sale of vermicompost grew by 500% in 2005. Vermitechnology Unlimited has doubled its business every year since 1991 (NCSU 1997; Kangmin, 1998).

US scientists are also searching for life-saving 'vermi-medicines' from the bioactive compounds in earthworms. (Mihara et al., 1990).

2) CANADA

Canada is also ahead in vermicomposting business on commercial scale for both 'vermicompost' and 'vermimeal' production. Large-scale vermicomposting plants have been installed at several places to vermicompost municipal and farm wastes and their use in agriculture (GEORG 2004). An 'Organic Agriculture Centre of Canada' has been established whose objective is to replace 'Chemical Agriculture' by widespread use of 'vermicompost' (Munroe 2007).

3) UK

UK is also following US and Canada in promoting vermicomposting mainly for waste management and to reduce the needs of 'waste landfills'. Large 1000 metric ton vermi-composting plants have been erected in Wales to compost diverse organic wastes (Frederickson 2000).

4) FRANCE

France is also promoting vermicomposting on commercial scale to manage all its MSW and reduce the needs of landfills. About 20 tons of mixed household wastes are being vermi-composted everyday using 1000 to 2000 million red tiger worms (*Eisenia andrei*) (Visvanathan et al. 2005).

5) NEW ZEALAND

It is also a leading nation in vermicomposting business. The Envirofert Company of New Zealand is vermicomposting thousands of tons of green waste every year. They put the green waste first to a lengthy thermophilic cooking, and then to vermicomposting by worms after cooling. Cooking of green waste help destroy the weeds and pathogens which may come from the feces of pets in grasses. They claim that each worm eat the cooked green waste at least 8 times leaving an end product rich in key minerals, plant growth hormones, enzymes, and beneficial soil microbes. Envirofert is also planning to vermicompost approximately 40,000 tones of food wastes from homes, restaurants and food processing industries every year. (www.envirofert.co.nz) (Frederickson 2000; Gary 2009).

6) PHILIPPINES

Vermiculture and vermicomposting were introduced in the Philippines in the 1970s. Vermicompost is being used by farmers on large scale replacing the chemical fertilizers. Recently, commercial production of 'vermimeal' from earthworms biomass has been started as a substitute to 'fishmeal' for promoting fishery industries. (Guerrero and Guerrero 2005).

7) ARGENTINA

Vermiculture is an expanding business in Argentina especially for the development of country-sides. 'Worms Argentina' is a growing company which reports to be exporting 'composting worms' on large scales to European, South American, Caribbean and Middle East nations. They are in high demands from Middle East countries for recycling of polluting dairy effluents. (Pajon, 2009).

8) CHINA

Vermiculture is a fast growing industry in China for the development of rural communities. Earthworms are being used for vermicomposting of 'waste', promoting 'organic farming' and for the development of 'vermi-medicines' and 'nutritive vermimeals'. (Sun 2003; Lopez 2003).

9) RUSSIA

Vermiculture is being promoted on large scale in Russia especially for the development of life-saving 'vermi-medicines' for treatment of human diseases for which conventional medicine do not have an answer. Scientists have developed a special breed of the versatile species *Eisenia fetida* which can tolerate and survive in cold climates. (Titov and Anokhin 2005).

10) JAPAN

Japan is promoting vermiculture since 1970s mainly for production of 'vermi-medicines' from the 'bioactive compounds' isolated from earthworms. (Tanaka and Nakata 1974; Wang 2000).

11) ITALY

In Italy, vermiculture is used to biodegrade municipal and paper mill sludge. Aerobic and anaerobic sludge are mixed and aerated for more than 15 days and in 5000 cum of sludge 5 kg of earthworms are added. In about 8 months the hazardous sludge is converted into nutritive vermicompost (Ceccanti and Masciandaro, 1999).

12) CUBA

Large scale vermicomposting of waste is going on in Cuba since the last ten years to produce an alternative to chemical fertilizer. In 2003, an estimated one million tons of vermicompost was produced (Munroe, 2007).

Vermiculture Movement in Australia: Diverting Wastes from Landfills

Vermiculture is being practised and propagated on large scale in Australia as a part of the 'Urban Agriculture Development Program' which utilizes the urban solid wastes. Australia's 'Worm Grower Association' is the largest in world with more than 1200 members (Dynes, 2003). The Sydney Waters in New South Wales has set up a vermiculture plant of 40 million worms to degrade up to 200 ton of urban wastes a week. Sydney's St. George Hospital is setting up plant to biodegrade its kitchen waste and fertilise its hospital gardens. The Gayndah Shire Council in Queensland, Australia, is vermi-composting over 600 tons of organic waste into valuable organic fertilizer (vermi-compost) and selling to the local farmers. This fertilizer is completely non-toxic and odour free. Other rural council in QLD Murgon Shire has also mounted a successful vermi-composting program.

Vermicomposting of dried sludge (biosolids) from the sewage and water treatment plants are being increasingly practiced in Australia and as a result it is saving large landfill spaces every year (Sinha et al., 2009 b). The Redland Shire in Queensland started vermi-composting of·sludge (biosolids) from sewage and water treatment plants with the aid of Vermitech Pty. Ltd. in 1997. The facility receives 400-500 tons of sludge every week with 17% average solid contents and over 200 tons of vermicast is produced every week by vermicomposting (Vermitech, 1998). The Hobart City Council in Tasmania, vermi-compost about 66 cum of biosolids every week, along with green mulch into 44 cum of vermicompost diverting them from landfills (Datar et al., 1997).

Vital Earth Company is vermicomposting waste on commercial scale. They have produced Solar Powered Vermiconverter VCS2000 and VCS4000. VCS200 has capacity of 2.2 cubic meter waste degradation producing 300 kg finished product (vermicompost) every week while VCS4000 has 3.5 cum capacity producing 420 kg of vermicompost every week. There is also production of 200 L of vermiwash (nutritive pesticidal liquid). Solar Powered Vermicomposters are ideal for schools, restaurants and other commercial organizations.

Important works on vermiculture and its significance in environmental management is going on at Murdoch University, WA, University of Western Sydney, NSW, Southern Cross University, NSW, University of Queensland, Brisbane, and Griffith University, Brisbane, QLD. (UNSW-ROU, 2002 a & b).

Vermiculture Movement in India: Converting Waste into Wealth (Brown Gold -'Vermicompost' & Green Gold – 'Food Crops') & Combating Poverty

Bhawalkar Earthworm Research Institute (BERI) in Pune, is most pioneer organization involved in vermiculture (vermicomposting of solid wastes) in India since the 1970s (Bhawalkar, 1995). The Delhi based Tata Energy Research Institute (TERI) is working for the cause of environment and propagating the 'vermicomposting technology' (VCT) for management of municipal solid wastes in India (Bhiday, 1995). Several agricultural universities are now involved in vermiculture studies and a movement is going across the sub-continent especially involving the poor rural women with dual objectives of 'making wealth from waste' while cleaning the environment. Earthworms have enhanced the lives of poor in India. In several Indian villages NGO's are freely distributing cement tanks and 1000 worms and encouraging women to collect waste from villages, vermicompost and sell to the farmers. It has become good source of livelihood for many. The production cost of vermicompost is estimated at Rs. 1.40 per kg in India and is being sold at Rs. 3.00 per kg. The cost of 1 kg of worms (about 2000 worms in number) is Rs. 1000.00 (about AU $30) in India (Senapati, 1992; White, 1997; Hati, 2001; Visvanathan et al., 2005).

Bihar, Karnataka, Tamil Nadu, Gujarat and Mahrashtra are leading states in vermiculture revolution. The Karnataka Compost Development Corporation established a first vermicomposting unit in the country to handle all municipal urban solid wastes and is producing 150 to 200 tons of vermicompost every day from city garbage. (Kale 2005). She has listed several farmers whose life has been changed from a poor 'farm labourer' to a 'rich farmer' who embraced vermiculture.

In the State of Bihar in India several farmers are being motivated to practice vermicomposting of their farm and food wastes and use the vermicompost as a nutritive bio-fertilizer for agriculture. Even educated unemployed in Bihar have now taken to vermicomposting business on commercial scale using all kinds of food and farm wastes as raw materials. A number of villages in the districts of Samastipur, Hazipur and Nalanda in Bihar have been designated as 'BIO-VILLAGES' where the farmers have completely given up the use of chemical fertilizers for the last four years since 2005. They are successfully growing cereals (rice, wheat & corn), fruits (banana, guava, mango & lemons) and vegetable crops (potato, tomato, onion, brinjal, cucumber, okra etc) on vermicompost at a much lower cost. Farmers of bio-villages feel proud of their food products and they sell at a higher price in market due to their good appearance and taste (Personal Interview by Rajiv Sinha, December, 2008).

Chapter 11

PROBLEMS ENCOUNTERED DURING VERMI-COMPOSTING & THEIR SOLUTIONS

1. POSSIBILITY OF UNPLEASANT ODOUR IN THE INITIAL STAGES

Although the system is generally 'odourless' as the worms secrete 'anti-pathogenic' coelomic fluid and also continually aerate the waste pile by their burrowing actions there may be initial odour problem in any vermi-composting process. This is mainly because the worms are overloaded with waste beyond its 'carrying capacity' at a given time and oxygen supply becomes insufficient leading to anaerobic conditions. Lesser number of worms per kg of waste biomass means less discharge of anti-pathogenic coelomic fluid. But as the worms grow mature and multiplies in numbers it discharges more fluid, creates more aerobic conditions in the waste biomass by burrowing actions, and also devour the anaerobic microbes responsible for rotting and odour. However, it is also advisable to manually turn and aerate the compost pile time to time and also add some carbon rich wastes (e.g. saw dust or dry grasses) to restore proper C/N ratio in waste materials if any odour problem occurs because this may be due to excessive nitrogenous wastes in the pile.

Researches indicate that a natural insoluble mineral zeolite (aluminum silicate) when mixed with waste (3-5%) reduces or eliminate the foul odor by absorbing the gases ammonia and hydrogen sulfide. Zeolite has additional advantages in vermi-composting. It has cage-like skeletal structure that allows to trap heavy metals which cannot then leach into environment to be extracted by plant roots (from vermicompost) or bio-accumulated by earthworms. Zeolite also helps to raise or lower pH of waste through cation exchange.

2. FRUIT FLIES, SOLDIER FLY MAGGOTS, MAGGOTS AND MITES

Fresh food scraps attract fruit flies and mites that cause some nuisance. The vermicomposting bins can also have an influx of 'soldier fly maggots'. Sprinkling limestone powder or putting some vinegar in a cup in the bin drives them away. Actual maggots may also appear initially if the organic wastes contain 'meat products', but it disappear with time as the population of worms grows and disinfect the system by discharging more coelomic fluid.

Chapter 12

CONCLUSIONS AND REMARKS

Our study has conclusively proved that management of 'food and garden wastes' by vermicomposting using waste eater earthworms done either methodically or even casually (as many people do not have enough time), can handle variety of such wastes much more efficiently than the conventional aerobic and anaerobic composting systems. Among the composting worms *Eisinea fetida* appears to be voracious & versatile waste eaters and decomposers which and is adapted to survive in both tropical and temperate conditions all over the world. However, worms have more preferences for the 'nitrogenous' food waste over the 'cellulosic' garden wastes (grass clippings).

Among the conventional composting methods, the 'anaerobic system' performed much better over the 'aerobic system' in both methodical as well as in casual ways of composting. It was contrary to the belief that 'aerobes' multiply & decompose waste faster than the 'anaerobes'. The greater efficiency of anaerobic systems can be attributed to overall 'better moisture & temperature regime' maintained in the composting pile. But, the negative aspect is that it emits high volumes of GHG.

In our vermicomposting studies (Valani, 2009) positive effects of application of 'lime' can be clearly seen in their 'improved efficiency' of degradation and increased values of NPK and beneficial microbes thus supporting the findings of Pramanik et al., (2007).

The nutritional and microbial studies of vermicompost indicates its 'superior & safer compost quality' over the two conventional composts made by aerobic and anaerobic systems. Several authors have also acclaimed the superior agronomic values of vermicompost over the conventional composts.

Our studies (Agarwal, 1999; Sinha et al., 2009 c) made on cereal and vegetable crops have also proved it.

Vermicomposting technology (VCT) is being scientifically improved worldwide now and is being commercialized all over the world for mid-to-large scale vermicomposting of most organic wastes (food & farm wastes & green wastes, the waste organics from various industries, and also the water & wastewater treatment plants sludge) and several companies have come up in the last few years. (Sherman, 2000; GEORG, 2004). Amending the various types of organic wastes from domestic, commercial and industrial sources, with 'cattle dung or cow manure' and application of 'lime' has been found to enhance the process of vermicomposting and produce a better quality of end-product. (Loehr et al.,. 1984; Dominguez, 2004; Tripathi & Bhardwaj, 2004; Pramanik et al., 2007; Muthukumaravel et al., 2008 & Valani, 2009).The worms 'love to feed on cattle dung' and any waste organics amended with cattle dung becomes its 'loved food'. Worms get a combination of pre-digested cellulosic & protein food along with large amount of microbes from the cattle dung.

Vermiculture is a growing industry not only for managing waste very economically but also for promoting allied industries depending on earthworms & vermicompost production e.g. organic farming & sustainable agriculture, fishery, poultry and dairy industries. It is estimated that one (1) ton of earthworm biomass on an average contain one (1) million worms approximately. One million worms doubling every two months can become 64 million worms at the end of the year. Considering that each adult worm (particularly *Eisinia fetida*) consume waste organics equivalent to its own body weight everyday, 64 million worms (weighing 64 tons) would consume 64 tons of waste everyday and produce 30-32 tons of vermicompost per day at 40-50% conversion rate.

Earthworms not only converts 'waste' into 'wealth', it itself also becomes a valuable asset as 'worm biomass' which are finding new uses for production of some 'life-saving medicines' and 'nutritive feed materials' besides their traditional uses in farms for improving soil fertility and enhancing crop productivity. Potentially large quantities of worm biomass will be available as 'pro-biotic' food for the cattle and fish farming, after the first year of composting.

Vermicomposting by waste eater earthworms should be encouraged in society as a 'home composting' program and among the commercial institutions (restaurants and groceries) dealing with food products to divert more and more 'food wastes' from ending up in the landfills which are 'economic and environmental burdens' on nations. Report says that every year 'food wastes' worth the value of $6 billion are thrown away in Australia from homes and commercial institutions, all ending up in the landfills. They can all be recycled back into 'nutritive fertilizer' through vermicomposting program.

Our study (Sinha & Chan, 2009) indicates that Brisbane residents who are 'home composting' their 'food & garden wastes' by any of the three methods of composting are diverting a significant amount of municipal solid wastes (MSW) from ending up in the landfills. A family of four (2+2) was diverting on average 4-5 kg of food & garden wastes every week which means at least 200 kg or 2 quintals of waste every year saving significant landfill spaces and reducing substantial financial burden on the local government. This also means those practising 'aerobic and vermicomposting of wastes' at home are also reducing emission of significant amounts of greenhouse gases.

As anaerobic composting systems emits the highest amount of CO_2, CH_4 and N_2O and vermicomposting system emits lowest amounts of all these GHGs and since the global warming potential of CH_4 and N_2O is approximately 21 times and 310 times that of CO_2, respectively, therefore on the basis of CO_2-equivalent, due to the 'extreme anaerobic conditions' prevailing in closed MSW landfills (all wastes are compacted and covered by soil everyday during filling) they would emit highest amount of GHGs (3640 mg CO_2-e/m^2/hour).

As more than 70% of the MSW is disposed off in landfills in Australia and in all developed nations, commercial vermicomposting of MSW can play a good part in the strategy of greenhouse gas reduction and mitigation in the disposal of global MSW (Toms et al. 1995; Sinha and Chan, 2009).

REFERENCES & ADDITIONAL READINGS

ABS (2004): Waste Management Services Australia – 2002-03; Cat No. 8698.0; *Australian Bureau of Statistics*, Canberra.

AGO (2007): *National greenhouse gas inventory 2005*; Report of Australian Greenhouse Office, Canberra.

Agarwal, Sunita (1999): *Study of Vermicomposting of Domestic Waste and the Effects of Vermicompost on Growth of Some Vegetable Crops*; Ph.D Thesis Awarded by University of Rajasthan, Jaipur, India. (Supervisor: Rajiv K. Sinha)

Appelhof, Mary (1997): *Worms Eat My Garbage*; 2nd (ed); Flower Press, Kalamazoo, Michigan, U.S. (http://www.wormwoman.com).

Appelhof, Mary (2003): Notable Bits; In *WormEzine*, Vol. 2 (5): May 2003 (Available at (http://www.wormwoman.com).

Atiyeh, R.M., Dominguez, J., Sobler, S. and Edwards, C.A., (2000): Changes in biochemical properties of cow manure during processing by earthworms (*Eisenia andrei*) and the effects on seedling growth; *Pedobiologia*; Vol. 44: pp. 709–724.

Bajsa, O., Nair, J., Mathew K. & Ho, G.E. (2004): Pathogen Die-Off in Vermicomposting Process; Paper presented at the International Conference on 'Small Water and Wastewater Treatment Systems', Perth, Australia.

Bansal, S. and Kapoor, K.K. (2000): Vermicomposting of crop residues and cattle dung with *Eisenia foetida*; *J. of Bioresource Technology*; Vol. 73: pp. 95–98.

Beetz, Alice (1999): *Worms for Composting (Vermicomposting)*; ATTRA-National Sustainable Agriculture Information Service, Livestock Technical Note, June 1999.

Bhawalkar, U. S. (1995): *Vermiculture Eco-technology*; Publication of Bhawalkar Earthworm Research Institute (BERI), Pune, India.

Bhiday, M.H. (1995): *Wealth from Waste: Vermiculturing*; Publication of Tata Energy Research Institute (TERI), New Delhi, India; ISBN 81-85419-11-6

Bogdanov, Peter (2004): The Single Largest Producer of Vermicompost in World; In P. Bogdanov (Ed.), *'Casting Call'*, Vol. 9 (3), October 2004. (http://www.vermico.com)

Buchanan, M.A., E. Russell and S.D. Block, (1988): Chemical characterization and nitrogen mineralization potentials of vermicomposts derived from different organic wastes. In: Edwards, C.A. and E.F. Neuhauser (Eds.) *'Earthworms in Environmental and Waste Management'* ; S.P.B Acad. Publ., The Netherlands, pp: 231-239.

Ceccanti, B. and Masciandaro, G. (1999): Researchers study vermicomposting of municipal and paper mill sludges; *Biocycle Magazine*, (June), Italy.

Chan, Y.C., Rajiv K. Sinha and W. J. Wang (2010): Emission of Greenhouse Gases from Home Aerobic Composting, Anaerobic Digestion and Vermicomposting of Household Wastes in Brisbane (Australia); J. of Waste Management & Research; UK. (http://www.sagepub.com). (Accepted)

Chauhan, Krunal (2009): *A Comprehensive Study of Vermiculture Technology : Potential for its Application in Solid Waste & Wastewater Management, Soil Remediation & Fertility Improvement for Increased Crop Production*; Report of 40 CP Honours Project for the Partial Fulfillment of Master of Environmental Engineering Degree; Griffith University, Australia (Supervisors: Dr. Rajiv K. Sinha & Dr. Sunil Heart).

Collier, J (1978): Use of Earthworms in Sludge Lagoons; In: R. Hartenstein (ed.) *'Utilization of Soil Organisms in Sludge Management'*; Virginia. USA; pp.133-137.

Contreras-Ramos, S.M., Escamilla-Silva, E.M. and Dendooven, L. (2005): Vermicomposting of Biosolids With Cow Manure and Wheat Straw; *Biological Fertility of Soils*, Vol. 41; pp. 190-198.

Dash, M.C (1978): Role of Earthworms in the Decomposer System; In: J.S. Singh and B. Gopal (eds.) *Glimpses of Ecology*; India International Scientific Publication, New Delhi, pp.399-406.

Datar, M.T., Rao, M.N. & Reddy, S. (1997): Vermicomposting: A Technological Option for Solid Waste Management; *J. of Solid Waste Technology and Management,* Vol.24 (2); pp.89- 93

Dominguez, J., Edwards,C.A , Subler, S. (1997): A comparison of vermicomposting and composting; *BioCycle* ; Vol. 28: pp. 57 -59.

Domínguez, J. (2004): State of the Art and New Perspectives on Vermicomposting Research; In C.A. Edwards (Ed.) '*Earthworm Ecology'*; pp. 401–424. CRC Press; Boca Raton, FL, U.S.A.

Dynes, R.A. (2003): *EARTHWORMS*; Technology Info to Enable the Development of Earthworm Production; Rural Industries Research and Development Corporation (RIRDC), Govt. of Australia, Canberra, ACT.

Eastman, B.R. (1999): Achieving Pathogen Stabilization Using Vermicomposting; *Biocycle,* pp. 62-64; (Also on Worm World Inc. Available www.gnv.fdt.net/reference/index.html. Seen on 24.7.2001).

Eastman, B.R., Kane, P.N., Edwards, C.A., Trytek, L., Gunadi, B., A.L. and Mobley, J.R. (2001): The Effectiveness of Vermiculture in Human Pathogen Reduction for USEPA Biosolids Stabilization; *J. of Compost Science and Utilization*; Vol. 9 (1): pp. 38 – 41.

Edwards, C.A. and Lofty, J.R. (1972): *Biology of Earthworms*; Chapman & Hall, London, 283 p.

Edwards, C.A., Burrows, I., Fletcher, K.E. and Jones, B.A. (1985): The Use of Earthworms for Composting Farm Wastes; In JKR Gasser (Ed.) *Composting Agricultural and Other Wastes*; Elsevier, London & New York, pp. 229 – 241.

Edwards, C.A. and Fletcher, K.E. (1988): Interaction Between Earthworms and Micro-organisms in Organic Matter Breakdown; *Agriculture Ecosystems and Environment*; Vol. 24, pp. 235-247.

Edwards, C.A. (1988): Breakdown of Animal, Vegetable and Industrial Organic Wastes by Earthworms; In C.A. Edward, E.F. Neuhauser (ed). '*Earthworms in Waste and Environmental Management'*; pp. 21-32; SPB Academic Publishing, The Hague, The Netherlands; ISBN 90-5103-017-7

Edwards, C.A. and Bohlen, P.J. (1996): Biology and Ecology of Earthworms (3rd Ed.), Chapman and Hall, London, U.K.

Edwards, C.A. (1998): The Use of Earthworms in the Breakdown and Management of Organic Wastes; In C.A. Edwards (Ed.), *Earthworm Ecology*; CRC Press, Boca Raton, FL, USA, pp. 327 – 354.

Edward, C.A. (2000): *Potential of Vermicomposting for Processing and Upgrading Organic Waste*; Ohio State University, Ohio, U.S.

Elvira, C., Sampedro, L., Benitez, E., Nogales, R. (1998): Vermicomposting from Sludges from Paper Mills and Dairy Industries with *Elsenia anderi*: A Pilot Scale Study; *J. of Bioresource Technology*, Vol. 63; pp. 205-211.

Fraser-Quick, G. (2002): Vermiculture – A Sustainable Total Waste Management Solution; *What's New in Waste Management* ? Vol. 4, No.6; pp. 13-16.

Frederickson, J. Butt, K. R., Morris, R.M., & Daniel C. (1997) : Combining Vermiculture With Traditional Green Waste Composting Systems; *J. of Soil Biology and Biochemistry*, Vol. 29; pp. 725-730.

Frederickson, J. (2000): The Worm's Turn; *Waste Management Magazine*; August, UK.

Garg, P., Gupta, A. and Satya, S., (2006): Vermicomposting of different types of waste using *Eisenia foetida*: A comparative study; *Bioresource Technology*, Vol. 97: pp. 391-395.

Gary, M (2009): Personal communication from Envirofert, New Zealand (gary@envirofert.co.nz)

Gaur, A.C. and Singh, G. (1995): Recycling of rural and urban wastes through conventional composting and vermicomposting; In: Tandon, H.L.S. (Ed.), *Recycling of Crop, Animal, Human and Industrial Waste in Agriculture*; Fertilizer Development and Consultation Organisation, New Delhi, India; pp. 31–49.

GEORG (2004): *Feasibility of Developing the Organic and Transitional Farm Market for Processing Municipal and Farm Organic Wastes Using Large - Scale Vermicomposting*; Pub. Of Good Earth Organic Resources Group, Halifax, Nova Scotia, Canada. (Available on http://www.alternativeorganic.com)

Graff, O. (1981): Preliminary experiment of vermicomposting of different waste materials using *Eudrilus eugeniae* Kingberg; In: M. Appelhof (ed.) Proc. of the workshop on '*Role of Earthworms in the Stabilization of Organic Residues*'; Malanazoo Pub. Michigan, USA.; pp.179-191.

Guerrero, R. (2005): *Commercial vermimeal production*; In Guerrero R, and Guerrero M (Eds.) *Vermitechnologies for Developing Countries*; Proceedings of the International Symposium on '*Vermi Technologies for Developing Countries*'; Philippines; p. 175.

Gunathilagraj, K and T, Ravignanam (1996): Vermicomposting of Sericultural Wastes; *Madras Agricultural Journal*; Coimbatore, India; pp. 455-457.

Hand, P (1988) : *Earthworm Biotechnology;* In : Greenshields, R. (ed.) *Resources and Application of Biotechnology : The New Wave*; MacMillan Press Ltd. US.

Hartenstein, R. and Bisesi, M.S. (1989) : Use of Earthworm Biotechnology for the Management of Effluents from Intensively Housed Livestock; *Outlook On Agriculture*; USA, 18; pp. 72-76.

Hati, Daksha (2001): 1000 Wriggling Worms and Rural Women; *The Deccan Herald*, 26th June, 2001, India.

Ireland, M.P. (1983): Heavy Metals Uptake in Earthworms; *Earthworm Ecology*; Chapman & Hall, London.

Kale, R.D., Seenappa S.N. and Rao J. (1993): *Sugar factory refuse for the production of vermicompost and worm biomass;* V International Symposium on Earthworms; Ohio University, USA. ·

Kale, R.D and Sunitha, N.S. (1995): Efficiency of Earthworms (*E. eugeniae*) in Converting the Solid Waste from Aromatic Oil Extraction Industry into Vermicompost; *Journal of IAEM*; Vol. 22 (1); pp. 267-269.

Kale, R.D. (1998): Earthworms : Nature's Gift for Utilization of Organic Wastes; In C.A. Edward (ed). *'Earthworm Ecology'*; St. Lucie Press, NY, ISBN 1-884015-74-376.

Kale, R.D. (2005) : *The role of earthworms and research on vermiculture in India*; In Guerrero R, and Guerrero M (Eds.) *Vermitechnologies for Developing Countries*; Proceedings of the International Symposium on '*Vermi Technologies for Developing Countries*'; Philippines; pp. 66-88.

Kangmin, Li (1998) Earthworm case: 4th ZERI World Congress; Windhoel, Namibia. (Also in Vermiculture Industry in Circular Economy; *Worm Digest* (2005) (http:// www.wormdigest.org/content/view/135/2/)

Kaushik, P. and Garg, V.K. (2004): Dynamics of biological and chemical parameters during vermicomposting of solid textile mill sludges mixed with cow dung and agricultural residues; *J. of Bioresource Technology*;, Vol. 4: pp. 203–209.

Kaviraj & S. Sharma (2003): Municipal Solid Waste Management Through Vermicomposting Employing Exotic & Local Species of Earthworms; *Journal of Bioresource Technology*; Vol. 90: pp. 169-173.

Klein, J., Hughes R.J., Nair, J., Anda, M. and G.E. Ho (2005): *Increasing the quality and value of biosolids compost through vermicomposting*; Paper presented at ASPIRE Asia Pacific Regional Conference on Water and Wastewater, Singapore, 10-15 July 2005.

Kristiana, R., Nair, J., Anda, M., and Mathew, K., (2005): Monitoring of the process of composting of kitchen waste in an institutional scale worm farm; *Water Science and Technology;* Vol. 51 (10): pp. 171-177.

Lakshmi, B.L. and Vizaylakshmi, G.S. (2000): Vermicomposting of Sugar Factory Filter Pressmud Using African Earthworms Species (*Eudrillus eugeniae*); *Journal of Pollution Research*; Vol. 19 (3): pp. 481-483.

Lisk, D.J. (1991): Environmental effects of landfills; *The Science of the Total Environment*; Vol. 100: pp. 415-468

Loehr, R.C., Martin, J.H., Neuhauser, E.F. and Malecki, M.R. (1984): *Waste Management Using Earthworms – Engineering and Scientific Relationships*; Project Report ISP – 8016764; National Science Foundation, Washington D.C.

Lopez, M (2003): *The use of earthworms as source of enzyme for medicines and for solid waste management*; Paper presented at the 1st Philippine Vermi-Symposium-Workshop; PCAMRD, Los Banos, Laguna, Philippines.

Lopez, M. and Alis, R. (2005): *Indigenous use of native earthworms and its fatty acids profile*; Paper Presented at the Int. Symposium on 'Vermitechnologies for Developing Countries'.

Lotzof, M. (2000): Vermiculture: An Australian Technology Success Story; *Waste Management Magazine*; February 2000, Australia.

Lou, X.F. and Nair, J. (2009): The impact of landfilling and composting on greenhouse gas emissions – a review; *J. of Bioresource Technology*; Vol. 100: pp. 3792-3798.

Mihara, H., Sumi, M., Mizumoto, H., Yoneta, T., Ikeda, R., & Maruyama, M. (1990): *Oral administration of earthworm powder as possible thrombolytic therapy*; Recent Advances in Thrombosis and Fibrinolysis; Academic Press, NY: pp. 287-298.

Mitchell, M.J., Horner, S.G., and Abrams, B.L. (1980): Decomposition of Sewerage Sludge in Drying Beds and the Potential Role of the Earthworm *Eisenia fetida*; *Journal of Environmental Quality*, Vol. 9; pp. 373-378.

Morgan, M., Burrows, I., (1982): *Earthworms/Microorganisms interactions*; Rothamsted Exp. Stn. Rep.

Munroe, Glenn (2007): *Manual of On-farm Vermicomposting and Vermiculture*; Pub. Of Organic Agriculture Centre of Canada; 39 p.

Muthukumaravel K., A. Amsath and M. Sukumaran (2008): Vermicomposting of Vegetable Wastes Using Cow Dung; *E-Journal of Chemistry*, Vol. 5 (4), pp. 810-813.

Nair, Jaya., Vanja Sekiozoic and Martin, Anda (2006): Effect of pre-composting on vermicomposting of kitchen waste, *Journal of Bioresource Technology*, 97(16):2091-2095.

Nair, Jaya., Kuruvilla Mathew and Goen, Ho (2007): *Earthworms and composting worms- Basics towards composting applications*; Paper at 'Water for All Life- A Decentralised Infrastructure for a Sustainable Future'; March 12-14, 2007, Marriott Waterfront Hotel, Baltimore, USA.

NCSU (1997): *Large Scale Vermi-composting Operations – Data from Vermi-cycle Organics,* Inc.; North Carolina State University, U.S.

Pajon, S. (2009) Vermiculture technology – Worms Argentina; Personal Communication (silvio.pajon@wormsargentina.com)

Parvaresh, A. et al., (2004): Vermistabilization of Municipal Wastewater Sludge With *E.fetida*; *Iranian J. of Environmental Health, Science and Engineering*; Vol. 1(2): pp. 43-50.

Patil, Swapnil (2005): *Vermicomposting of Fast Food Waste*; 20 CP Project submitted for the partial fulfilment of the degree of Master in Environmental Engineering; School of Environmental Engineering, Griffith University, Brisbane; November 2005. (Supervisor: Rajiv K. Sinha).

Pierre, V, Phillip, R. Margnerite, L. and Pierrette, C. (1982): Anti-bacterial activity of the haemolytic system from the earthworms *Eisinia foetida andrei*; *Invertebrate Pathology,* 40, pp. 21-27.

Pramanik P., G.K. Ghosh, P.K. Ghosal and P. Banik (2007): Changes in organic – C, N, P and K and enzyme activities in vermicompost of biodegradable organic wastes under liming and microbial inoculants; *J. of Bioresource Technology*; Vol. 98: pp. 2485-2494.

Reinecke, A.J., Viljoen, S.A., and Saayman R.J. (1992): The Suitability of *Eudrilus eugeniae, Perionyx excavatus* and *Eisinia fetida* for Vermicomposting in Southern Africa in Terms of Their Temperature Requirements; *J. of Soil Biology and Biochemistry*, Vol. 24: pp. 1295 – 1307.

Sabine, J.R. (1978): *The nutritive value of earthworm meal*; In Proc. Of the Conf.on '*Utilization of Soil Organisms in Sludge Management*'(Ed. R. Hartenstein); Syracuse, NY; pp. 122-130.

Satchell, J. E. (1983) : *Earthworm Ecology- From Darwin to Vermiculture;* Chapman and Hall Ltd., London; pp.1-5.

Saxena, M., Chauhan, A., and Asokan, P. (1998): Flyash Vemicompost from Non-friendly Organic Wastes; *Pollution Research*, Vol.17, No. 1; pp. 5-11.

Seenappa, S.N. and Kale, R. (1993): Efficiency of earthworm *Eudrillus eugeniae* in converting the solid wastes from the aromatic oil extraction units into vermicompost; *Journal of IAEM*; Vol. 22; pp.267-269.

Seenappa, S.N., Rao, J. and Kale, R. (1995): Conversion of distillery wastes into organic manure by earthworm *Eudrillus euginae; Journal of IAEM*; Vol. 22; No.1; pp.244-246.

Senapati, B.K., (1992): Vermitechnology: An option for Recycling Cellulosic Waste in India. In: *New Trends in Biotechnology*. Oxford and IBH Publications Pvt. Co. Ltd. Calcutta., pp: 347-358.

Sherman, Rhonda (2000): Commercial Systems & Latest Development in Mid-to-Large Scale Vermicomposting; *Biocycle*; November 2000, pp. 51.

Singleton, D.R., Hendrix, B.F., Coleman, D.C., Whitemann, W.B. (2003): Identification of uncultured bacteria tightly associated with the intestine of the earthworms *Lumricus rubellus; Soil Biology and Biochemistry*; Vol. 35: pp. 1547-1555.

Sinha, Rajiv K. (2000): *Waste Management : The 3 R's Philosophy* ; INA Shree Publisher, India; ISBN 81-86653-32-5; 318 p.

Sinha, Rajiv. K., Sunil Herat, Sunita Agarwal, Ravi Asadi, & Emilio Carretero (2002): Vermiculture Technology for Environmental Management: Study of Action of Earthworms *Elsinia foetida, Eudrilus euginae* and *Perionyx excavatus* on Biodegradation of Some Community Wastes in India and Australia; *The Environmentalist*, U.K., Vol. 22, No.2. June, 2002; pp. 261 – 268.

Sinha, Rajiv K., Sunil Herat, P.D. Bapat, Chandni Desai, Atul Panchi & Swapnil Patil (2005): Domestic Waste - The Problem That Piles Up for the Society: Vermiculture the Solution; *Proceedings of International Conference on 'Waste-The Social Context*; May 11-14, 2005, Edmonton, Alberta, Canada; pp. 55-62.

Sinha, Rajiv K. and Rohit Sinha (2007). *Environmental Biotechnology* (Role of Plants, Animals and Microbes in Environmental Management and Sustainable Development); Aavishkar Publisher, Jaipur, India; ISBN 978-81-7910-229-9; 315 p.

Sinha, Rajiv K. & Gokul Bharambe (2008): Vermiculture for Sustainable Solid Waste Management : Making Wealth from Waste While Diverting Huge Organics from the Landfills and Reducing Greenhouse Gases; *Indian Journal of Environmental Protection (IJEP)*, ISSN 0253-7141; Regd. No. R.N. 40280/83; Indian Institute of Technology, BHU, India.

Sinha, Rajiv K, Gokul Bharambe & David Ryan (2008 a):Converting Wasteland into Wonderland By Earthworms: A Low-Cost Nature's Technology for Soil Remediation : A Case Study of Vermiremediation of PAH Contaminated Soil; *The Environmentalist*; UK; Vol. 28: pp. 466 – 475; Published Online 14 May 2008, Springer, USA.

Sinha, Rajiv K, Jaya Nair, Gokul Bharambe, Swapnil Patil & P.D. Bapat (2008 b): Vermiculture Revolution : A Low-Cost & Sustainable Technology for Management of Municipal & Industrial Organic Wastes

(Solid & Liquid) by Earthworms With Significantly Low Greenhouse Gas Emissions ; In James I. Daven and Robert N. Klein (Eds): *Progress in Waste Management Research*; NOVA Science Publishers, NY, USA ; pp. 159 - 227; ISBN : 978 – 1 – 60456 – 235 – 4.

Sinha, Rajiv K., Sunil Herat, Gokul Bharambe, Swapnil Patil, P.D. Bapat, Kunal Chauhan & Dalsukh Valani (2009 a): Vermiculture Biotechnology: The Emerging Cost-effective and Sustainable Technology of the 21st Century for Multiple Uses from Waste & Land Management to Safe & Sustained Food Production; *Environmental Research Journal;* Vol. 3 (Issue 1); pp. 41-110; NOVA Science Publishers, NY, USA.

Sinha, Rajiv K. Sunil Herat, Gokul Bharambe & Ashish Brahambhatt (2009 b): Vermistabilization of Sewage Sludge (Biosolids) by Earthworms: Converting a Potential Biohazard Destined for Landfill Disposal into a Pathogen Free, Nutritive & Safe Bio-fertilizer for Farms; *J. of Waste Management & Research;* (Published On-line August 26, 2009) (http://www.sagepub.com).

Sinha, Rajiv K. and Andrew Chan (2009 c): Study of Emission of Greenhouse Gases by Brisbane Households Practicing Different Methods of Composting of Food & Garden Wastes : Aerobic, Anaerobic and Vermicomposting'; NRMA – Griffith University Project (Report Submitted to GU & NRMA).

Sinha, Rajiv K., Dalsukh Valani, Sunil Herat, Krunal Chauhan & Kulbaivab Singh (2009 d): Vermitechnology for Sustainable Solid Waste Management: A Comparative Study of Vermicomposting of Food & Green Wastes With Conventional Composting Systems to Evaluate the Efficiency of Earthworms in Sustainable Waste Management With Reduction in Greenhouse Gas Emissions; In Justin A Daniels (Ed.) Advances in Environmental Research - Vol. 9 (Chapter 3); NOVA Science Publishers, N.Y., USA; ISBN: 978 – 1 – 61728 – 999 – 6.

Sinha, Rajiv K. (2010): *Vermiculture Revolution : The Technological Revival of Charles Darwin's Unheralded Soldiers of Mankind*; Book; NOVA Science Publishers, USA. (Under Publication)

Singh, Kulbaivab (2009): Microbial and Nutritional Analysis of Vermicompost, Aerobic and Anaerobic Compost; 40 CP Honours Project for Master in Environmental Engineering; Griffith University, Brisbane, Australia; (Supervisors: Dr. Rajiv K. Sinha & Dr. Sunil Heart).

Standard Australia (2003): Australian Standard (4454 – 2003) for Compost, Soil Conditioners and Mulches; Pub. Of Australian Standard Bureau, Canberra.

Suhane, R.K. (2007): *Vermicompost* (In Hindi); Pub. Of Rajendra Agriculture University, Pusa, Bihar; 88 p. (www.kvksmp.org) (Email : info@kvksmp.org)

Sun, Z-J (2003): *Vermiculture and vermi protein*; China Agricultural University Press; Beijing, PRC; 366 p.

Suthar, Surindra (2009): Vermicomposting of Vegetable-Market Solid Waste Using *Eisenia fetida*: Impact of Bulking Material on Earthworm Growth and Decomposition Rate; *Ecological Engineering*.

Tanaka, B. and Nakata, S. (1974): Studies of 'antipyretic components' from the Japanese earthworm; *Tokyo Igaku Zasshi*; Vol. 29: pp. 67-97.

Titov, I.N. and Anokhin B.M. (2005): The ten-year results of treatment with the extract of earthworm tissues; Innovation Centre, Moscow, Russia; In Guerrero R and Guerrero M (eds.) *Vermitechnologies for Developing Countries*; Proceedings of the International Symposium on '*Vermi Technologies for Developing Countries*'; Philippines; pp. 148-149. (titov@green-pik.ru)

Toms, P. Leskiw, J., and Hettiaratchi, P. (1995): Greenhouse Gas Offsets : An Opportunity for Composting; Presentation at the 88th Annual Meeting and Exhibition, June 8-12, 1995, San Antonio, Texas, USA; pp. 18-23.

Tripathi, G., and Bhardwaj, P. (2004): Decomposition of Kitchen Waste Amended With Cow Manure Using Epigeic Species (*Eisinia fetida*) and an Anecic Species (*Lampito mauritii*); *J. of Bioresource Technology*; Vol. 92: pp. 215 – 218.

UNSW, ROU (2002 a): *Vermiculture in Organics Management- The Truth Revealed*; (Seminar in March 2002) University of New South Wales Recycling Organics Unit; Sydney, NSW, Australia;

UNSW, ROU (2002 b): *Best Practice Guidelines to Managing On-Site Vermiculture Technologies*; University of New South Wales Recycling Organics Unit; Sydney, NSW, Australia; (Viewed in December 2004) www.resource.nsw.gov.au/data/Vermiculture%20BPG.pdf

Valani, Dalsukh (2009): *Study of Aerobic, Anaerobic & Vermicomposting Systems for Food & Garden Wastes and the Agronomic Impacts of Composts on Corn & Wheat Crops;* Report of 40 CP Honours Project for the Partial Fulfillment of Master of Environmental Engineering Degree; Griffith University, Australia (Supervisors: Dr. Rajiv K. Sinha & Dr. Sunil Herat).

Vermitech (1998): Successful Biosolids Beneficiation With Vermitech's Large-Scale Commercial Vermiculture Facility in Redlands; *Waste Disposal and Water Management in Australia*; Vol. 25 (5); September-October, 1998.

Visvanathan et al. (Eds.), (2005): *Vermicomposting as an Eco-tool in Sustainable Solid Waste Management*; Asian Institute of Technology, Anna University, India.

Wang, Y.S., Odle, WSI, Eleazer, W.E., and Baralaz, M.A (1997): Methane Potential of Food Waste and Anaerobic Toxicity of Leachate Produced During Food Waste Decomposition; *Journal of Waste Management and Research*; Vol. 15: pp. 149-167.

Wang, Z.W. (2000): Research advances in earthworms bioengineering technology; *Medica*;Vol. 31(5); pp. 386-389

Wengling, C. and Sun J. (2000): *Pharmaceutical value and uses of earthworms;* Vermillenium Abstracts; Flowerfield Enterprizes, Kalamazoo, MI (USA).

White, S. (1997): A Vermi-adventure in India; *J. of Worm Digest*; Vol. 15 (1): pp. 27-30.

Wu, XL., Kong, HN, Mizuochi, M., Inamori, Y., Huang, X., and Qian, Y. (1995): Nitrous Oxide Emission from Microorganisms; *Japanese Journal of Treatment Biology*, Vol. 31 (3): pp. 151-160.

Xing M., Yang, J. and Lu, Z. (2005): *Microorganism-earthworm Integrated Biological Treatment Process – A Sewage Treatment Option for Rural Settlements;* ICID 21st European Regional Conference, 15-19 May 2005; Frankfurt; Viewed on 18 April 2006. <www.zalf.de/icid/ICID_ERC2005/HTML/ERC2005PDF/Topic_1/Xing.pdf>

Yaowu, He., Yuhei, Inamori., Motoyuki, Mizuochi., Hainan, Kong., Norio, Iwami., and Tieheng, Sun (2000): Measurements of N_2O and CH_4 from Aerated Composting of Food Waste; *J.of the Science of The Total Environment*, Elsevier, Vol. 254: pp. 65-74.

INDEX